THE CREATING COSMOS

Books by Barbara Dewey

The Creating Cosmos
The Theory of Laminated Spacetime
As You Believe

The Creating Cosmos

by **Barbara Dewey**

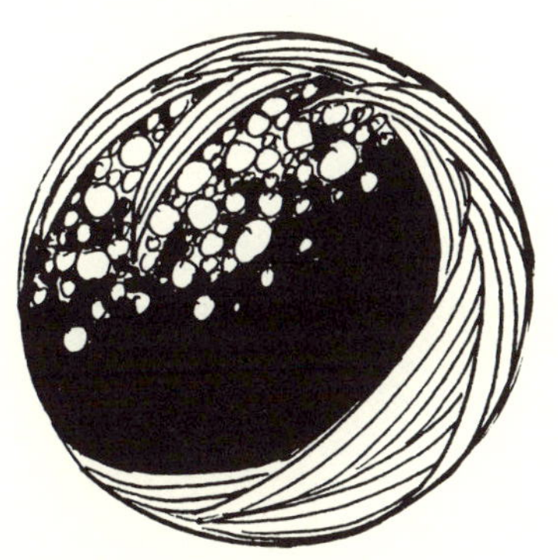

BARTHOLOMEW BOOKS
INVERNESS, CALIFORNIA
1985

Have you come to accept the sacredness of All That Is?
Then this book is dedicated to you.

Contents

1.

A New Answer to Old Questions

WE HAVE BEEN IN SEARCH of answers to why things happen as they do ever since Adam woke to find one of his ribs missing and a naked lady staring at him dolefully. The answers we have found credible may have little to do with the facts concerning any particular matter but they have certainly had much to do with the way we have lived our lives, because answers—if they satisfy our sense of logic—become Truths. And Truths become inviolate guides in the pursuit of our goals.

Down through time, some of the deeper philosophical questions we have asked are these:

•What is the purpose of life?

•Is there a God?

•If there is a God, why does He put us through such suffering?

•If there is no God why are we here?

•How was the universe created? Why?

•Is there some purpose to its design?

•What are the laws which govern the human condition?

There is more to these musings than mere idle speculation. We need a context within which we can make appropriate choices. For instance, if there is a God, and this God is demanding that we follow The Ten Commandments, then we will know that when we don't, we will get into trouble. Conversely, if there is no God, then following The Ten Commandments may only be an exercise in contrived self-righteousness to which we need not subscribe. Answers, then, are vital to the decision making process which weaves the fabric of our lives.

We have moved, over the centuries, from religious answers which explained the cosmos as the creation of a God to scientific answers which explain the cosmos as the chemical consequence of a Big Bang. Traditionally, Western religion has described the earth as a place where the children of a judgmental God have been put to receive their rewards and punishments for their moral behavior. It is therefore inferred that the abiding intent of God is to whip us into spiritually acceptable shape if it takes disease, famine, and flood to get the job done. Given such a truth, it may be that we are insufficient in the eyes of our Creator but at least we matter. We matter so much that we can enrage The Almighty.

Science scrapped these truths and developed its own. We were spawned, science says, not in the Garden of Eden, but in primordial slime. The purpose—at least of animal life—is to improve the species by killing off the biologically inferior members—a kind of eating-one's-way-to-survival program. While science officially offers no judgment with its conclusions, we cannot escape the philosophical inferences of such a theory; we matter little except as biological curiosities. We play out our time in the Theatre of the Absurd. The most worthy purpose to which we can aspire under the auspices of scientific theory is to keep killing off the weak and helpless so that the dross may be sifted from the fine. As there is war and other forms of mayhem going on all over the world at this very minute, never should it be said that we are not living up to what science believes to be nature's best hope for us.

Something else happened to us when we switched from religious truths to scientific truths. In the religious concept of the cosmos we were central to its importance. It was designed with us in mind. The sun travelled our heavens to offer us warmth and light. The sky cupped the earth in a loving embrace. The scientific mock-up of the cosmos is very different. We become little more than mold on a wet speck of dust, caught in a solar system which is only one of billions. Solar systems come and go as accidentally as the light of a firefly. And when our sun goes, this curious experiment will all be done with. Given such a proposition, it's hard to believe it matters, in this age of scientific wisdom, whether or not we mount a nuclear Armageddon, or treat each other with kindness. Science, perhaps unwittingly, has raised the milieu of accident and meaninglessness to the level of the exalted. In our celebration of science and its truths, we have, of necessity, become existentialists, coping as best we can with an ordered but purposeless world. We have, in so far as we have changed from traditional religious truths to scientific truths, switched from the fear of damnation by a not-so-benign father to deliverance into an orphanage.

*   *   *

Perhaps it was a dissatisfaction with the truths of both religion and science which took me on a different intellectual path. Or perhaps it was just sheer perversity which refused to accept the wisdom of others as my own. In any event, about thirty years ago, I set out upon an independent path. Certainly I was not in search of anything as august as a cosmic design. All I wanted was to know why life hurt so much and why I was so miserable. In an effort to get at the cause of such misery, I underwent psychoanalysis.

Hundreds of sessions and two divorces later I was only a little better off psychologically, but I had, in the process, developed a keen interest in the mind. I began to read copiously about it and the more I read, the deeper I wandered into the field of psychic experience. One book, in particular, was to change my life dramatically. It was *Science of Mind* by Ernest Holmes. In it, Holmes described our individual minds as being an indivisible part of what he called Universal Mind. He likened the relationship to a drop of water in an ocean except this particular ocean, he said, is without time and without space. It is an infinite, eternal creating medium which is an integral and indivisible part of all. Drop a thought into this creating medium in the right way and the thought will *have* to manifest physically. Holmes was saying, "Thoughts create things," just as Jesus had, although Jesus said it this way: "As ye believe, so shall it be done to thee." It was hard to credit but *if* it were true, then a whole new world could open up before me.

With considerable more doubt than conviction, I began to test Holmes' statements in very small ways. And he was right! I *could* create what I wished! To make a trial and error adventure short, I eventually learned to use the power of my mind to create "healings"—healings for money, healings for cancer, healings in all kinds of directions. There seemed to be only two requirements. My intent had to be strong and my faith in the outcome complete. I must also disregard the concepts of time and space. This meant that I must produce *now* in the present an image or thought so strong, so unambivalent, that it would *have* to manifest physically in what we think of as the future.

As a side benefit of even greater importance, I came to see through these experiments with the use of mind that I was indeed the sole creator of the patterns of my life in every way. The sorrows that I was heaping upon myself were nothing but the out-picturings of a mind in the negative mode. I was "healing" for negative results by dwelling on negative thoughts using exactly the same laws as I used for positive healings. The only difference was that I was unconscious of one act and conscious of the other. Universal Mind didn't seem to care what I created, but as long as I used my mind to imagine, I was going to reap what I sowed. (My book, *As You Believe,* details the entire healing process. I describe it as a self-help book for those who want to improve the quality of their lives.)

So it was that I was thoroughly familiar with and charmed by this spaceless, timeless creating medium called Mind when my father[1], a scientist of some repute, came West for a visit. In a moment of idle conversation he told me of a theory he had about the character of time, a subject which had always interested us both. It was a concept which totally shattered all scientific concepts about the mechanics of the cosmos. He called it Laminated Spacetime (dis-continuous time). Exactly what he meant by that and the inferences which such a hypothesis suggests are the subject of this book. For the moment it is sufficient to say that a cosmic design of immense significance fell into place for me.

I did what I could to tell my father of my new insights but he turned a deaf ear. I was being too metaphysical, he grumbled. I rather think that he, so late in life, could not afford to find cause within a subjective reality where my experiments with mind suggested cause lay. Such a truth would have confronted too harshly his wish for a mathematically objective universe. He remained too much the servant of his truths, in spite of his great insight. At any rate, I kept my thoughts to myself after several rebuffs.

That was many years ago and the wheel has turned considerably since then. We, as a civilization, are floundering more noticeably in pessimism and fear. We drive ourselves more inexorably towards nuclear annihilation in, I believe, a symbolic statement of our suicidal despair. Even scientists, our current high priests, are beginning to doubt the base upon which they have founded our contemporary religion. It has therefore become more urgent to share good news if there is such to share. And indeed, I do have good news. Lots of it.

Before we get into the meat of the matter, however, I would like to give you some idea of why our beliefs, our truths, are so important to the course of our lives. I would hope that you would then have a new appreciation of the effect your own truths may be having on your own life. Beliefs (truths) are far more important to the

---

[1]Edward R. Dewey. Founder of the Foundation for the Study of Cycles, fellow (deceased) of the World Academy of Arts and Sciences, recipient of a gold medal from the Biometric Research Foundation.

conduct of life than friends, for beliefs not only select our friends for us, but the very conditions under which we force ourselves to live.

I also want to discuss some aspects of our inherited wisdom which most of us take for granted. In this I have an ulterior motive. I want to do what I can to discredit much of that wisdom because so much of it is dangerous to a sense of well-being and self-worth. I think it could be well argued that if a truth diminishes us as humans, it is probably not a truth at all but a distortion. (Running like an ugly undercurrent through my mind as I write that statement is the popular belief that we, as humans, are the kings of the jungle and entitled therefore to have our way with the rest of nature. I consider such arrogance to be in itself a diminishment of our human value. Surely our capacities are greater than are demonstrated by brute force and/or cunning.) I want to make room for *new truths*. And, to rephrase Daniel J. Boorstin, ''The greatest obstacle to discovery is not ignorance, but the illusion of knowledge.'' So, I want to shatter, if I can, the illusions which bind us to our present knowledge as if it were Truth.

My father's theory—The Theory of Laminated Spacetime—is a theory which deals with physics. I have developed the physical aspects of this theory in a book entitled *The Theory of Laminated Spacetime*. I recommend it to anyone who is interested in physics and/or the design of the physical cosmos. I might add that it is written *by a layman for laymen* and explains the *causes* of gravity, magnetism, polarity, and much more. I have no doubt that it will be taken seriously and will eventually become the basis for tomorrow's physics. When this happens—in the way of circular logic—the theory will prove itself and then its subjective implications, or what I have come to call the Creating Cosmos, will be given even greater legitimacy. But you, dear heart, can be years ahead of the game if you are willing to accept such ''proof'' on faith and your own good judgement.

The Theory of Laminated Spacetime isn't half so important to our cultural progress as are the subjective implications of that theory. Theories about physics neither detract from, nor improve upon, let us say, the effects of gravity. They may delight the minds of the curious, but they do nothing directly to change the basic physical laws by which we live. They serve only to elucidate and, of course, they may open up new technological paths, perish the thought. (I think of Einstein and the atomic bomb.)

But the Creating Cosmos which sprang from the Theory of Laminated Spacetime is very different from a theory which deals only with physical laws. It could change for all time the beliefs we hold about ourselves, the solutions we apply to our problems, and our relationships to our greater global environment. It could end forever the current misconception that there is a separation between the spiritual and the physical. Along with reordering physics, the concept of the Creating Cosmos might even reorder religion. Such a change could very substantially change the quality of life for us all. And that, I think, *is* important.

2.

Truths

TRUTHS ARE BELIEFS which have had tabernacles built around them. Let it be said at the outset, however, that all truths are man-made. As such, all truths are subject to fallibility, and should be held in tentative friendship, not in wedlock. Truths are the way we catalogue what little wisdom we dare lay claim to. Yes, we avidly search for truth but the quest is—and should be—endless. The truths we find along the way speak more to a description of our path than to any hypothetical place of destination.

Fortunately, we do not need to know truth in the absolute in order to use it as it is intended. I say fortunately because we are demonstrably shy of any wisdom which effectively guarantees safe passage through life's suffering. If we had to wait upon our arrival at wisdom's door before we could experience some success in that venture we would likely all be done for. What we do need, however, is the self-assurance which *believing* we have hold of a truth inspires. Then armed, if only with the illusion of right guidance, we can relax in the psychological calm of imagined certitude, safely beyond de-stabilizing doubt.

We humans, unlike most animals, do not accept stoically the discomforts of being cold or hungry or sick or bereft of those we love. We have small tolerance for physical pain. We categorically reject the emotional distress of being wrong or guilty. Nor can we support the agony of being found unworthy or rejected. Given the spectrum of life's possibilities and the small portion of that spectrum we are willing to embrace enthusiastically, it is little wonder that we have given a great deal of attention to investigating the causes of our few successes and our many failures. Nor need we be surprised that we mark so well the paths which divide themselves into right action (pleasure) and wrong action (pain). The sign posts we label as Truths are supposed to offer guidance away from what we don't like and towards the happiness we so fervently seek.

Coping with the pleasure/pain principle is only one aspect of the tasks we assign to our truths. We live within a complicated fabric, woven as much by other people and other events as by ourselves. Forces, seemingly beyond our control, either block us from our goals or—with some luck—carry us closer to them. Fortune, presumably, necessitates working tediously through a maze of actions and inter-actions, all of which have a rippling and multiple number of consequences. So that we may stay on top of this jumble, our truths must not only point the way to success, but they must also attempt to make sense of what would otherwise appear to be patently unreasonable.

Forced to make minute by minute decisions—all of which we ardently hope are the correct ones—it becomes absolutely essential to make those decisions within a meaningful context. We need the Big Picture in which to carry out our small plans. Our truths, then, must provide some sort of cosmological frame of reference for us. When we ask ourselves why it is that we live in such a loony-bin, the answer which explains life as we witness it, must make it all appear reasonable. Our truths, then, serve to rescue us from the disorganization which uncertainty inflicts and knowingness distances.

Until the advent of science as the Ultimate Purveyor of Wisdom. supplying this cosmological context, these answers to why some kinds of action produced disastrous results and some kinds produced success was the province of religion. In ignorance or omniscience, religion was the keeper of the flame which lit the path to truth and right action and hence the explanation for, if not the avoidance of, suffering. The responsibility of religion was to delineate the context in which life took place, to explain what the abiding purpose of life was, and accordingly, to interpret where right action lay as well as the promises it offered. Now these prerogatives have been largely usurped by science because we have come to believe that it and it alone deals with facts (truths).

*   *   *

Any truth which purports to be such should account for circumstances as they are, not circumstances as one might wish they were. Both science and religion have been almost fanatical in disregarding this prime requirement of any truth. Religions deal only with the subjective world of spirit when it pronounces upon right action. If they had their way, we would not be physical beings at all. Science deals only with the physical aspects of life, failing to acknowledge any importance the soul might have on that physical life. Thus the priest deplores the sick as sinners and the scientist/doctor orders them to take two asprin and go to bed. Neither cares, then, for the whole man. Each treats only his half of that person according to his particular prejudices.

When religion was in full control of the lives of its faithful, it asked the devout to turn the pleasure/pain principle upside down. The faithful were expected to deny themselves the comfort and pleasures of their bodies as well as all ego satisfaction. It was intrinsically unappealing. It invited a challenge. And the challenge finally came from science which did not hold with the truth that discomfort was part of the grand design. Willingly we followed such a promise. But, when our scientists became our high priests of wisdom, we were also forced to accept their cosmos and their truths—truths which denied our importance and our spiritual component. It was a choice which has had a profound effect upon our self-confidence, but to understand why we must look again at what we require and how our truths affect these requirements.

Second only to our need for a context within which to judge right and wrong action is our need to have cause for hope. We live on hope. However vaguely articulated, something within us yearns for the accomplishment of an inspired goal. We want the realization of the dreams that our imaginations invent. Beyond the wish for physical safety and comfort, what we dream of is the expression of our special talents in such a way that we are hugely appreciated for that specialness. We wish to live in our idea of a perfect world and be credited for our contributions towards that perfection. We want to matter—to make a difference—and to know that we do. It is the hope of this achievement which empowers the will to live, for if we believe that it doesn't matter one way or the other then it makes very little sense to endure the hardships of living. These hopes are spiritual hopes. They nourish the soul. And only a nourished soul can keep the body in the game of life.

It is not so much that we must accomplish these goals. Rather, we must believe that such goals are possible. For that we need self-confidence and the strength which self-confidence inspires. If this confidence is threatened by self-doubt or if our truths diminish our sense of worth, then hope, the flame behind the will to live, is undermined and the paralysis of fear can set in. Survival not only becomes increasingly difficult, but the very wish to survive is also diminished as hope diminishes. Primarily, survival does not depend on being biologically fit but on being *psychologically* fit and self-assured. Truly, man does not live by bread so much as he lives by dreams.

Our truths directly determine what kind of dreams, if any, we permit ourselves and how we view ourselves as part of those dreams. If we hold, as a truth, the notion that a critical God is watching our every mistake, intent upon punishing us for those mistakes, then our dreams of personal worth are likely to be substantially diminished. How can we think well of ourselves when God's love is conditioned upon a pass/fail test we are more than likely to fail? With such truths we are apt to describe the world as dangerous and life as difficult as we thread our way between right and wrong. We are also likely to view with revulsion those who do not give the same careful attention to the violation of these truths. In short, we are apt to hate others as much as we hate ourselves. Such feelings are not conducive to a sense of well being. They breed war, an option we can ill afford in the Nuclear Age.

On the other hand, if we believe that the universe is a mathematical construction of which we are but a pin-point on a speck of dust swirling in one small galaxy and that our genesis as a latter-day phenomenon is more of a curiosity than a marvel, we are likely to view the world about us quite differently. Instead of every action having its own, albeit frightening, importance, *any* action, particularly if it is taken in the ironical name of personal effectiveness, will seem quite absurd. We become philosophic nihilists. "What the hell difference does it make?" we must ask. And the answer is implied in the truth. "None!" I do not think it an accident that two thirds of our most brilliantly sophisticated minds—our scientists—now work directly or indirectly for the Pentagon.

Truths, then, are far more important to us as psychological frames of reference than for any intrinsic piece of wisdom they might or might not contain. Indirectly they imply reason for hope or they invite despair. Truths also establish our priorities and the kind of world those priorities will establish. Today for every 100,000 people in the world there are 556 soldiers and 85 doctors. The cost of one nuclear submarine is equivalent to the cost of educating 160 *million* Third World children for one year. We prefer the submarine.

*   *   *

Religions have traditionally asked much of us and the road to salvation (the realization of worth) is usually the unpleasant one of self-denial and sacrifice. Major religions tend to view God's handiwork as unconditionally perfect until he inexplicably suffered a bad day and, in some sort of colossal kitchen catastrophe, produced his only mistake—mankind. But if religions view us with repugnance, they also offer an outside chance of making good— not here, of course, where we could see if this truth were correct— but *somewhere,* in heaven or another life, depending upon one's truths. Thus, while deploring humans for their failures and very carefully refusing to blame God for his responsibility in that failed project, religions do make promises and thence offer hope, however dimly shines the light of that hope.

Compare the cosmic mock-up of religion with Darwin's Theory of Evolution, the most widely accepted Western explanation of our origins and purposes. Here science associates the will to live, not with a hope inspired by dreams, but with brute strength and biological luck. With the Darwinian Truth the body is all that matters. Dreams of satisfying the soul boil down to biological fitness and concerns for health, exercise, diet, cosmetics, and fat farms. Our dreams are of clothing fashions, sports idols and the body beautiful and its sexual proficiency.

Right action in the service of this truth sanctifies undisciplined and self-centered competition. Therefore it is not only egotistically satisfying to see our enemies vanquished, the very laws of nature demand that the weak perish on the battlefield, in poverty pockets, in the market place, and on our city streets. It is nature's will that the strong accomplish this process.

Religions may have found us seriously wanting, but at least we mattered. We mattered so much that we could even infuriate God. Now, if indeed there is a God, he is largely indifferent. We are no longer the children of God, but the grandchildren of apes. Admitted or not, this is cause for terrible feelings of loneliness and alienation, a malaise from which an increasing number of our population suffers. It can be seen in the increasing welfare rolls, the increasing need for medical attention, as well as the increasing dependence on street drugs which are sought by the rich and poor alike. (Money, it should be observed, does not cure hopelessness. In many ways it may even make it worse, for to chase a rainbow and find the pot at the end of it empty can be particularly disappointing.)

Somewhere behind the Big Bang and that pool of primordial slime there may still be a God in the official view of science, but if so his presence is certainly remote and his power problemetical. Moreover, optimism remains high that after a few more turns in the road, his hypothetical existence will be disproved once and for all.

It may be argued that people and their spiritual comfort are not and never were the province of science. Indirectly this is true, but science has seen fit to explain our origins and infer our purposes. They have dissected our livers and our cells. They purport to cure our illnesses and they pronounce upon our every need from fluoridated water to life-support systems. The scope of their so-called expertise as well as their methodology has seriously distorted all fields of inquiry. Yet nowhere in their articles of faith is there mention of ourselves, our spirit, our souls—that internal something which causes us to love and weep, to dream and despair. That part of us is ignored in their truths. It is therefore impossible to assume that this part of us matters or else it would have been accounted for in their theories. As this is the part of us we truly cherish—which truly matters to *us*—a belief in scientific truths forces the conclusion that we, *as humans,* don't matter at all.

So how come we abandoned God and replaced him with a scientist? To discover some of those answers we need to look more closely at both religion and science. I would like you to consider them for what I believe them to be—half truths—quite literally.

What is so unfortunate about these half truths is that they denigrate us. They offer so little in the way of hope and personal effectiveness. As a matter of fact I think it might easily be argued that if a truth has negative implications concerning our origins and purposes, it is not a truth at all but someone's jaundiced fantasy. The Creating Cosmos I wish to introduce you to (after we have explored our current myths) is positively alive with encouragement. We live in a universe which has purposely loaded the dice in our favor. We, with our passions and our dreams, *are* the universe.

3.

Religious Truths

ELIGION FILLS A DYNAMIC and urgent human need—the need each of us feels to be interconnected with the power, as well as the majesty, of life itself. Without these ties we feel alienated, powerless; we live in fear. In that fear, we turn to the false security of MX missiles and the position of overkill as a public policy. We need drugs—from aspirin to heroin—so that we may be protected from life itself. We cherish neither ourselves or others. From rape to political assassinations authorized by our government, we express contempt for life.

Religions combine into one focus an intuited knowledge that all life is sourced by a powerful wellspring and our sense of the ideal. Neither of these aspects of total reality is of the physical world. They are subjective or spiritual dimensions symbolized by our concept of a soul. Religion then tends, or should tend, the needs of the soul.

It was perhaps a natural part of the imaginative process to invent gods who both inhabited and managed this invisible world of the soul. It may also have been natural that these gods would display the ultimate characteristics we could, as humans, imagine. But when we took the best we could imagine in ourselves and externalized that into something which was not ourselves, but divine, we paid a heavy price. That which was ideal was divine;

that which was less than ideal (ourselves) was mortal.

Following the division between good (gods) and less-than-good (ourselves) we took a second step away from the truth. We ascribed opinions to these gods—human opinions, mind you, for to speak for any god is to speak only through human limitation. Gods became much more than perfected examples of the outer limits of human imagination; they became judgmental. Historically, by the time the many gods of pantheism became the one Judaic-Christian god of Western religions there was a schism of august proportions between God The Perfect (whose image we had invented) and humankind (whose self-assigned image of imperfection we now despised) because we imagined that he too despised us for those imperfections.

Within the context of a deified Ideal—where the Ideal has been objectified into a God without—we need to feel that union with that Best is possible. That is to say, if we are to externalize the best within us into an image called God, there must be some perceived means of achieving harmony with that God. We must be able to achieve a state of grace. We must feel we have a chance at achieving those dreams of goodness which keep us going. If we put that in terms of God, then we must be able to experience his approval. Unfortunately, traditional religions have devised such an obstacle course along this path to grace that success is all but impossible.

According to the orthodox religious view, what seems to raise God's hackles the quickest is the appetites. The very act of wanting something seems to be wrong. Yet ''wanting something'' is what provides the motivation for our goals and propels life forward. Historically, what religions would have us forego is any satisfaction or real pleasure in physical life except the twisted pleasure of self-denial. Instead of honoring life as a natural spiritual manifestation, they have held it in contempt. The faithful are encouraged to fear their natural impulses and to loathe themselves for experiencing them. We need, for instance, to enjoy sex, not only because of its reproductive importance but for its transcendent spiritual communion. Yet religions, in general, find sexual activity unwholesome except as it must be permitted to insure another generation.

It is my own suspicion that the Church found sex in particular and joy in general to be too much of a challenge to its power base. Unhappy people make better supplicants. The Shrine of Delight had, I think, to be destroyed through the insinuation of guilt if the masses were to be kept flocking to the Shrine of Devotion and Financial Sacrifice. These attitudes have been strategically beneficial to the Church but it has had its disastrous effects upon our cultures.

I must say it again. All religions are founded on human perceptions. They must therefore be described as man-made. However wise our religious teachers may have been, they can be but the vessels for truth, not truth itself. They must necessarily shape the truth as a vessel shapes its wine. It therefore must be understood that it is *we* who do not like ourselves, not God. It is *we* who are afraid of physical pleasure, not God. It is *we* who perceive ourselves as evil, not God. And it is *we* who make judgments upon human behavior, not God. God's judgment has already been imposed through natural law and the invention of the body.

*   *   *

Let's go back and look more closely at our religious history so that we may understand how our religious truths came to be and why they have, in so many ways, failed us. Whether we describe ourselves as atheist, agnostic, or believer, all of us are profoundly affected—for better or worse—by the culture in which we live and by that culture's religious roots. For Western culture, this means the Judaic-Christian theology of the Old and New Testaments.

The concept of one God found in the Old Testament represented a radical departure from what are now depreciatingly called primitive or pantheistic religions. I think of the religions of the Eskimos and our own Indians as good examples of pantheism. In these religions, divine spirit is within all physical manifestations. The spiritual path is based upon mutual respect and cooperation among all aspects of life. In many ways the Ecology Movement of today represents a return to these spiritual values. It might well be said that the best of religion and science come together here in a twentieth century setting.

Pagans did not see the spiritual world this way. They invented a super-structure of bigger-than-life people to rule the invisible universe and to account for the power source which lay beyond the physical world. The paganism of the early Greeks and Romans illustrates this model of religious belief. With paganism a dichotomy is established which breeds confrontation and competition. The threads which run between humans and their gods are webbings of jealousy, envy, manipulation, suspicion, etc. Roman paganism was the religious base upon which Christianity was imposed—by force. It became politically expedient to shift from paganism to Christianity and this could be accomplished most easily by making as few changes as possible.

I think it might be fair to say in the light of our history that when we anthropomorphize Spirit in our own image instead of appreciating Spirit in its myriad manifestations, we get the Crusades and acid rain. We get intolerance and the conflict of personalized self-interest. Our struggle with our gods is our struggle with the enemy. We think of ourselves as being at war with life, instead of in sacred communion with it.

Within the context of this kind of confrontational theism, the assumption is that our woes are attributable either to a sort of divine caprice or to some requirement in the process of give and take between ourselves and our benefactors which we failed to appreciate properly. As we shaped our gods in our own image and not the other way around, they were complete with our own weaknesses and lusts. We therefore conciliated them as we would wish to be conciliated—with bloody sacrifices and/or opulent offerings as a means of effecting their good humor.

We laugh now at the naivete which developed answers so wide of the mark and stuck by them through pestilence and famine. Surely there must have been some evidence that the sacrifice of a virgin or two didn't always guarantee the success of the community. Yet belief systems select their own evidence and force their own conclusions. Once established, they become very difficult to change for there is an imperative behind belief which disregards logic. They become Truths.

It is this characteristic of belief which makes our own times so dangerous. We know that war as we used to understand the term is no longer feasible. Even in conventional warfare, winners have been losers in all major twentieth century conflicts. Russia, for instance, lost twenty million lives, almost $3\frac{1}{2}$ times what the Jewish community lost in the extermination programs of Hitler. What did she win? The opportunity to rebuild a wasteland and fight the United States for world dominance. Personally, I don't think such goals are worth one life.

The concept of world dominance, incidentally, is a concept which comes into the heads of political leaders when they rise to the top of their own political heap. They like to keep going towards grander dreams. They are afraid of looking weak. They cast their own aspirations and fears in public terms and evoke patriotism so that they can use their country's young men to satisfy their own egos and assuage their own fears. This is what Hermann Goering, Hitler's deputy, had to say on the subject:

> "Of course people don't want war. Why should some poor slob on a farm want to risk his life in a war when the best he can get out of it is to come back to his farm in one piece? . . . it is the leaders of the country which determine the policy and it is always a simple matter to drag the people along, whether it is a democracy or a fascist dictatorship, or a par-

liament, or a communist dictatorship . . . All you have to do is tell them why they are being attacked and denounce the pacifists for their lack of patriotism and exposing the country to danger. It works the same in any country."

Whatever the specious causes of war, we still believe in war as a meaningful solution. It remains, for us, a truth. If we do not change this truth to something more reasonable before tempers run short or there is a computer accident, we could blow up the entire world. That certainly makes the sacrifice of a few virgins look pretty innocent, doesn't it? When we consider the million dollars *a minute* we spend internationally on armaments and the non-renewable resources we sacrifice on the altar of our belief in war—without so much as a temple to mark the expense unless you count the Pentagon—the question must arise as to who is being more absurd. Pagans or latter-day Christians?

At any rate, we in the West emerged from pantheism into a belief in one God. While the simplification may have forced a finer focus on the source of our difficulties, it left a yawning gap between ourselves and that one Supreme Being. There was one judge and no jury. Under paganism, at least it was always possible to find a divine ally. But how can we be sure that the one God is blessing *our* battlecamp?

*　*　*

Genesis presents what is now called the Creationist point of view, though for much of Western recorded history that point of view was orthodox Truth. It sets forth an explanation for the origins of life, the cosmic design into which life falls, and the reasons for human suffering. *Genesis* makes it abundantly clear that misfortune and life are so immutably intertwined that the two are nearly synonymous. Furthermore, it is within the intrinsic nature of man himself where the cause of tribulation can be found. *We are endowed with original sin!* We are intrinsically wicked from the very onset.

It seems self-evident to me that God had entrapment in mind when he designed the Garden of Eden and threw in the snake. Adam and Eve merely fell for the set-up. Even before they could gather up a few fig leaves, God was on the scene *in person*—for it was that sort of a disaster—purple with rage and vowing to get even with the couple if it took every forthcoming generation.

It was not an auspicious beginning!

I have to wonder at the ineptitude and/or intent of a God who would create such a seriously flawed pair to tend a world which had taken some effort to create and with which he had, so short a time ago, been well pleased. I wonder even more seriously about the authors of such a story and why they wished to invent such a hate-filled story. If one uses the apple incident to represent freedom of choice, then the authors of this part of the story were exceedingly wise to see that a ''God-given'' freedom gives us the liberty to succeed or suffer as we wish. But God, having given that freedom, should have rejoiced that this freedom was being acted upon. It therefore seems likely to me that a second group of authors, wishing to underscore a moral persuasion of their own, must have tacked on God's imprecations in order to drive home their own personal point of view.

Be that as it may, the total story was, for centuries, taken very seriously and subscribed to at its most superficial, narrative level. It still is. The clear implication is that the purpose of life is to survive in a booby-trapped and precarious world ruled by a captious tyrant. The Judaic-Christian religions are therefore based on the founding assumption that the Creator hated (and with very good reason) his own creations.

How differently the Greeks accounted for the world's sorrows. How lacking in moral judgment. Here too it was an act of disobedience, but it implied no endemic flaw within the human character nor any guilt to be atoned. Pandora opens a box and lets out all our troubles. Instead of being treated to scorn and loathing, however, she is helped to find something of a solution. She is asked to open the box once again. This time she lets out the Fairy of Hope—that necessary constituent of all life. Hope is allowed to struggle after the world's troubles and offer solace.

Hope, of course, is born of the possible, but I do not find that possible within the Old Testament. The Old Testament God is a nag—intemperate, overproud, and self-indulgent. His stock in trade is fear, not hope. If we are to appreciate properly our religious legacy we must feel deeply the psychological gravity of being loathed by, separated from, and in contention with, our deity. Such conditions produce spiritual anxiety and self-contempt of immense proportions.

At least Christians found in Jesus a sort of good fairy savior who could function as an intermediary. They also created a hierarchy of saints. The Jews incorporated no such mediators into their religion. As a consequence, they have had to go it alone. I'm not sure but that this has left them significantly troubled and psychologically disadvantaged. I personally feel that the Old Testament God is much too tempermental for anyone to handle. He can breed little but the feeling of congenital defeat within his subjects—and his followers *are* subjects. Whether or not this has proved a weakness and encouraged Jews to martyrdom, it has been manifestly demonstrated throughout history that the Jewish community has suffered much with recourse to little except a faith in stoicism.

Contrasted to the personal despair which the Old Testament brings to Jews and Christians alike, the gods of the Greeks and Romans rollicked their way through one hedonistic adventure after another. The only chasm which separated a man from his god was that a god had a greater capacity for misconduct than was possible for a mere mortal. Forgiveness, if it entered the thought process at all, must have been more of a human attribute than a divine one. In any event, this arrangement was most certainly egalitarian and it did little to damage a person's self-respect, for moral considerations were in their infancy in this part of the Mediterranean. In terms of personal pride, such a model might have served us well except that the Roman emperors took to proclaiming themselves gods in order to account for and excuse their excesses.

Christianity was dying the natural death of a cult when Emperor Constantine seized upon it as a device to enforce absolutism. It is my observation that a specific religion cannot flourish unless it receives state support and, of course, is prepared to pay the price by supporting the state. Be that as it may, Christianity was saved from obscurity because it could be used as a political tool.

The Old Testament supports the concept of the divine rule of kings and a judgmental and powerful God who could and would back up such divine law. This is, I suspect, why the Old Testament was incorporated into the Christian Bible. Constantine's lineage was not royal and he needed all the support he could find to back up his claim to his throne. It was, then, not so much what the Bible had to say on religious matters but its political attractiveness which rescued it from oblivion. Moreover, Christianity came to be championed by a politically and militarily strong papal state. That didn't hurt either. I do not mean to imply that there is little of spiritual value or beauty within the Bible. But I do mean to point out that its wisdom has survived in good measure because it was politically expedient that it do so. History does not record the rise and fall of other religions, other systems of wisdom, not similarly blessed though perhaps even more inspiring.

Among its other virtues, Christianity could also demonstrate to an overtaxed and war-ravaged population that its low station and life of hardship was all part of a Divine Design For Good. Christians had come to interpret meekness with spiritual grace. Such a pearl was a real gem to an emperor intent upon re-enforcing absolutism. When Jesus said, "The meek shall inherit the earth," meekness did not imply passiveness but rather something more akin to awareness—a cognitive understanding of man's rich interconnection with God-The-Power. But meek, as it came to be interpreted, was a quality of character any despot could admire—at least in others—and feel comfortable about encouraging.

With this kind of corrupting interpretation of the teachings of Jesus, everyone had to consider his soul above his bodily needs. (The wealthy, of course, managed to do both.) Fearing eternal damnation, but promised salvation if they but obeyed their priests and princes, the common folk were properly cowed and kings could be assured a comfortable rule over their subjects if not over competing kings. The Church enjoyed the new political game as much as did the princes. No doubt, it felt the word of God could be enunciated more convincingly if it were backed by large holdings, displays of great opulence, and an energetic army. And the church was right!

The good news was that sometimes the equality of political power between a particular king and the Church served to temper the excesses of both. Even a king had to consider his soul eventually—usually by buying the Church's approval with his army or his gold. As there was no judicial system worthy of the name, this stand-off provided the best the times could offer in the way of checks and balances. The bad news was that collusion between the Church and State became a political art form of blatant proportions.

*　*　*

There is another aspect of the Adam and Eve story which should be noted because it has had such a profound effect upon our history. Man's downfall is attributed to the female influence. The story would have been on far safer grounds intellectually if Adam had been born of Eve in the normal way. If this aroused fears of incest, then another birthing method could have been invented which at least took into account the natural, life-sponsoring capacity of the female half of the pair. By stigmatizing Eve and defining her as a troublemaker, our unconscious perceptions of life have been importantly distorted.

We still live very much within a culture where those qualities perceived as female qualities are held by both men and women to be suspect, or inferior, to the so-called male characteristics. Domination is better than cooperation, action is better than contemplation, reason is better than intuition. The world of things takes precedence over the world of spirit.

In terms of consequence, it is not so important that women *per se* have been made second class citizens. Far more telling is the fact that female qualities have been made into second class qualities and women have been kept from making that kind of contribution to the culture. Character traits like nurturing, subjectivity, and compassion have been suppressed within men because there was little nobility in such qualities. Our culture is based in large part on a masculine sense of honor, a masculine externalized religion, and a masculine interpretation of valid knowledge (science). Our history is a male history. *His* story, not *her* story. We have concentrated upon exploration, technology, and materialism—all aspects of an out-turned concentration. We place relatively little value on the arts, ethical considerations, a rich inner life—aspects of ourselves which are in-turned (female).

This concentration upon a masculine mode has importantly warped our sciences—"the art of procuring and evaluating knowledge." Molded to conform to the masculine requirements of "objectivity," all traces of female subjectivity have been consciously exorcized (or so it is hoped) from the fact-finding process. Oversimplified, this means that what is seen is taken to be true and what is not seen has no importance to the sciences. It means truths are arrived at through the statistical process, not the heart. Truths are the product of a laboratory and wisdom is gained through a test tube. But if the truth must consider *all* the parts, what does this do to scientific "facts?" Do not scientific truths necessarily become half truths when the female aspect of an indivisible whole is disregarded? Do we perhaps only have the words without the music to what science insists be called a song?

Unhappily, the thrust of the women's movement has not been towards validating what is female but rather in demanding that women be allowed to express themselves in a masculine mode. I vigorously support that right. But imitation still supports the erroneous notion that the masculine mode of expression is superior and we will not be significantly better off for that endorsement. There will just be more people on one side of a boat which is listing badly already.

*　　*　　*

There is little real correlation between the teachings of Jesus and the teachings espoused by the majority of Christian sects. Jesus preached brotherly love, a loving God, and the *art of self-emancipation*. Yet even in this religiously liberal country, the majority of clergy support both the concept and the act of war, the idea of a judgmental God and an insistence on obedience to ecclesiastical authority. What's more, Methodists can't even get along with Baptists. How on earth are they expected to get along with Moslems or even Jews?

Historically, the Church has usually been in the forefront of hate wars and it has always been divided by sectarianism. Our contemporary Holocaust was fired as much by the missionary zeal of Christian hate as it was by Aryan delusions of grandeur. Jews, always a favorite target for Christian antipathy, became the twentieth century Infidels, heretics, and pagans of other Christian out-reach programs in other centuries. I daresay if Hitler had been a Jew persecuting Christians, the Christian nations would not have been so indifferent to what was going on in pre-war Germany. We didn't start to get stuffy about his activities until he began to attack Christian nations.

It would be a great over-simplification to attribute Europe's past barbarisms solely to Christianity, but the fact remains that there is something about Christianity specifically which encourages self-righteousness—a self-righteousness which spills outwardly as hate. The abiding Christian urge seems to be to convert 'em and if you can't convert 'em, kill 'em.

Let's look at this self-righteousness/hate connection for a moment. Whatever one's personal beliefs, the entire world is now caught up in a fight to what may be the finish between two nations. I do not think it an accident that both have deep Christian roots. Today, in this country we call the Infidel, or pagan, a Communist. And the Russians call him a Capitalist. Very little else has changed. Both nations mean to convert the rest of the world or kill it. While officially Russia is atheistic, it remains, by heritage, deeply rooted in a Christian ethic which cannot be legislated away with any speed. Indeed, the Christian concept of a paternal God has been used unconsciously to support oppression in Russia for centuries. Unquestioned paternalism is the root assumption of both Christianity and totalitarianism.

The Christian ideal established nothing less than divine perfection as its goal. Christians were presented with an awesome challenge. They were expected to be bodiless, egoless, discarnate spirit and to avoid all temptation of physical pleasure and self-approval while living within a physical setting. The task set before aspiring Christians can be likened to a demand that a lily grow arms and legs if it is to be acceptable *as a lily*. Bodies and egos and the demands they make upon us are necessary to a soul in its physical state, yet these components were regarded with revulsion by Christian high-achievers. St. Catherine of Sienna, one of the most revered of all saints, spoke passionately of her "wretched filthiness." And it was this kind of talk which won her admiration and sainthood.

It is the same kind of loathing for the body which is at the base of the Right to Life Movement—the anti-abortion lobby. It is sexual license which disturbs the majority of its members, not the assault on incipient life. You do not find these same people in the fight against capital punishment, nuclear disarmament or other life-protecting pressure groups. Indeed, they are very likely to think our criminal justice system is soft on criminals, favor a strong military build-up, and encourage the ownership of handguns. Basically, punishment for wrong-doing is what they want because punishment for licentiousness is very much a part of the Christian ethic. Sex-for-pleasure is an intolerable license to many. It demands some sort of punishment, never mind the kind of life an unwanted child may have to endure. It is a thirst for revenge which rises from an internal pressure to do-unto-others-what-I-am-doing-to-myself.

The abhorrence of the necessary and natural packaging in which we must live forces the self to hate the self for what it is. This is the loathing which creates the spiritual anxiety to which I referred earlier. It is an agony founded on trying *not to be* what one *must* be. With such a canker in the heart, is it any wonder that there is a drive to make others play by the same rules? Self-loathing, outwardly expressed, becomes hate for others. Guilt, outwardly expressed, becomes the condemnation of others. Anxiety, outwardly expressed, becomes the fear of others. Thus, spiritual anxiety leads to a fanatical distrust of others because the self is so distrusted.

*　*　*

Religion is defined as a system of faith and worship. In a high-tech society where traditional religion has lost its compelling appeal, the religion—the system in which we place our faith and regard "with extreme respect"—becomes science. Therefore our contemporary religious leaders are scientists. Our latter-day priests may be even more powerful than their predecessors because they claim to offer pragmatic proof of their omniscience—proof we, as laymen, are unable to dispute. Because religious leaders, whatever their point of view, establish the procedures for living the attuned life within the culture they serve, now, instead of the confessional and prayer, we have annual medical check-ups and better living through chemistry.

The price we have paid for elevating science to the position of a religion is that we have been forced to live without spiritual hope in a context of impotence and fatalism. Science purposely severed all connection with the stuff of spirituality when it chose to focus on the physical world alone. This in itself is not necessarily an error, if the facts which arise from such an investigation are labeled accordingly. The problem is that scientists have long since forgotten their own "original sin"—the conscious division of an *in*divisible reality into two parts. One part scientists describe as objective and important to the gathering of knowledge. The other part they describe as subjective and detrimental to the gathering of knowledge. We are, whatever science has to say on the matter, 100% subjective—a point which will be taken up in the next chapter. This means that they have importantly distorted all their facts. They have also dismissed the very heart of life as unimportant to an understanding of life. This is why I say that they have forced us to live without spiritual hope if we live only with their truths.

Let's look at just one of their truths, a truth which is by no means pivotal in the scientific mock-up of the universe but which does nonetheless symbolize the kind of defeatism which runs through their truths—truths we have come to believe because they are pronounced by our priests. Scientists speak knowingly of the eventual disintegration of the sun. It is, they say, consuming itself in anorexic glee through the process of nuclear destruction. (This "fact" is disputed in *The Theory of Laminated Spacetime,* Chapter 4.) No matter how far off they say the time will be when the sun is no more, it makes the future finite and thus diminishes hope. It also negates any idea we might have had that we were provided for by a thoughtful God and thereby it doubly diminishes hope. All life is, in the final analysis, nothing but one of nature's minor little experiments, an aberration of chemical possibility but meaningless in the Big View.

I think scientists rather enjoy doing this to us; it serves us right for ever thinking we were important in the scheme of things. Scientists prefer to see us as the end product of a chemical chain of events which happened *to us* but to which we make no contribution. Scientists, in general, are not very comfortable with people. They prefer a laboratory or a notebook. I think it importantly affects how they view mankind and the degree of respect they have for humans.

As inadequate as we are in the eyes of traditional religion, at least our inadequacy is something we partially create for ourselves in wrong action. Right action can therefore effect salvation. By accepting the wisdom of science, it makes no difference what we do. The world is amoral, indifferent and going to end anyway. We serve no greater purpose than to endure in the pollution they have created for us and play with the bombs they have invented for our distraction while we wait for the world to end in one way or another. Do you wonder at the moral indifference reflected in our crime rate, or our preference for obliterating the experience of living with the overuse of drugs and alcohol?

Our new religion has managed to make the same sweetheart deal for itself as other successful religions. It too is state supported, no less than other religions in other countries in other centuries. This means that science can only maintain its position of power as long as it behaves supportively. Most research programs exist only because of government financing. This is not altruism on the part of the government. The government wants the kind of research which supports its own political needs and it relies heavily on the right kind of announcements and products coming from these high priests.

Scientists, on the other hand, prefer to do research rather than sell insurance. As long as they can produce direct political power for the government (new armaments, for instance) or create a direct need for more government (high cost medical technology comes to mind), scientists can keep their jobs.

Democratic governments derive their power from the support of the people. The weaker the people feel themselves to be, the more they will feel the need of a strong government and the greater will be their support for that kind of government. Governments instinctively like a dependent (docile) population. They like to foster price-support programs, welfare rolls, and trade restrictions which make the affected portions of the populations government-dependent.

Behind the military/industrial complex to which Eisenhower so ominously referred, is the political/scientific complex. And both elements of the complex look, perhaps unconsciously, to fear on the part of the citizenry to increase their particular authority. The government can extend its power if it can point to an enemy (in our case the Russians) from which we need protection. (There is an added benefit to finding an enemy beyond our borders; it unifies the country behind a common goal and makes governing easier.) Accordingly, scientists will be honored (and remunerated) as long as they can produce the instruments by which we can be saved from the threat which the political leadership has invented.

Science, too, relies upon a fearful population to maintain its authoritative position. We spend billions on technological health care—an estimated 90% of it for the last year of life—because we are in such fear of death. We demand high-speed transportation and communication systems because we are so afraid of losing the competitive race . . . to where? for what? We stock and re-stock our arsenal, truly not so much because we are afraid of Russian strength, but because we are afraid of our own innate weakness and our inability to manage in a computerized, technological, threatening world without God to help. Cultural fear is heady stuff for those who are sustained by their own roles as saviors.

There is therefore no need to look for a cheap or simple solution to any of our problems as long as the government and the new religion of science are in bed with each other. Self-interest of the colluding giants does not lie in that direction.

* * *

In the West, these are the traditional religious truths in which our culture is rooted. We are born inherently wicked; we are born blighted with original sin. Our bodies are neither to be trusted or cherished for they represent all that is ugly about the human condition, a condition which itself is a form of punishment for that original sin. God is judgmental; he punishes wrong-doing and rewards righteousness. If you doubt the depth of this truth, test yourself. How many times have you wondered why a particular fate befell a friend. "He was such a good man," you reason. "Why did it happen to him?" Behind this reasoning is a sense of divine punishment and reward. Punishment and reward even continue beyond the grave, imposed by an unforgiving, unrelenting God who will assign us either to heaven or to hell.

These truths combine to create a cultural belief system which gets in the way of self-approval and self-confidence, attitudes which a psychologically healthy person needs in his or her pursuit of a worthwhile life. Such truths establish both a feeling of deep betrayal by a God who would create such a flawed product and a terrible sense of personal inadequacy. They produce fear of wrong action and a low guilt threshold if wrong action does occur. They invite a sense of constantly being spied upon which produces disquietude.

I also think such truths as these breed criminal behavior. If a person once begins to see himself as bad or unworthy, he will, in most cases, give up and try to win his own particular race by being importantly bad, not just small-time bad. He or she becomes hate-obsessed. The day each of us is truly happy and proud of himself for who he is *as he is,* we can stop expressing ourselves in war or crime. Neither will we need life-obliterating drugs. Most illnesses will disappear.

Such detrimental effects as I have just described are the negative aspects of the legacy handed to us by institutionalized Western religion. This was not what Jesus taught, or rather, tried to teach. Jesus taught love for self and love for others. Little wonder we need a Second Coming. Clearly we were not listening the first time.

Even so, there is a brighter side to the influence of Christianity upon the lives of us all. If it is to be accused of sponsoring our infamies, credit must be given for the high state of truly noble inspiration it generated. The world may never again see such an outpouring of superlative art and music and architecture as flowered under the stimulation of Christian fervor.

It suggests to me that beyond the small-minded, self-serving dicta of the institutionalized Church, in a very real way Christianity also tapped into the exalted best of man and it did offer mankind hope, however distantly placed. While we have gained something for ourselves in the way of self-respect as the power of the Church waned, we have also lost Christianity's incalculable gift to us when we transferred our allegiance to science. Perhaps this gift was the capacity to love outrageously, unabashedly the source of life, to glimpse the majesty of possibility for mankind, and to feel a passion for something which transcended the self—all made possible through the promise of hope elevated to cosmic proportions. Without this hope, of which we presently have so little, our own historical legacy to those who come after will not be a cathedral but a toxic waste dump.

4.

Scientific Truths

T HE CHALLENGE FOR SCIENTISTS, as it was for religious philos-
ophers, was also the question *why?* But their whys were small
whys so their answers were not truths but facts. Why does a ball
bounce? Why is blood red? Why does an apple fall? It was not
their purpose to still an anxious heart but to satisfy a curious
mind.

Scientists were looking for knowledge. The ambiguities of faith
and opinion distressed them. "Let us begin," they said, "by
scraping away all that is not universally verifiable. Let us cleanse
from the truth-gathering process all matters of faith, emotion,
perceptual bias, opinion. Let us deal only with objective facts, or
what can be demonstrated predictably and observed dispassion-
ately. Only if a fact has been divested of all its subjective possibilities can we be certain it is really a fact. Conversely, if it does not
fulfill these requirements, we cannot call it true."

It was a noble undertaking, but the good men failed to understand two fallacies in their procedures which would deny them
their goals. First, they themselves could never escape their own
subjectivity. And second, nature is a synthesis of both the seen and
the unseen, the physical and the non-physical. (The "objective"
and the subjective.) It does not serve the cause of truth to sever the
two parts of nature and find all cause and all effect in one half of

35

that complementary whole. We cannot study physical matter by itself if we hope to understand the laws of nature because nature is both the physical and the non-physical.

Let's look at the first misconception.

It is an irony that when you ask a question, select your data, and determine the approach to that data, you have already made three subjective choices which will prejudice your findings accordingly. For instance, it seems obvious to us that water always runs downhill and so we might well ask why. But this is a very limited point of view. Stand off a distance from the planet and we might be asking why water ran every-which-way—up, down, sideways. Observe the planet from an even greater distance and the question of water might never arise. Were we alien to this planet, we might find that water was not even visible. It might seem like air-currents and we might only think to ask why it eddied. The best we can say is that when we do speak of water, we, the observers, bring so much bias to our observations that the questions we ask and answer speak more to our own special idiosyncrasies than they do to any definitive truth.

We like to think that our senses—our eyes in particular—bring us objective information. If we see it with our eyes, it exists. This is the principle criterion by which we judge physical reality. But our eyes are programmed to see only a very small band of light waves. Change our eyes ever so slightly and much of what we see now would be lost to us and much of what we don't see would become visible. The line between what we call real and what we think of as not real could change dramatically. What we call real is then a phantom reality, forced upon us because our eyes "lie." This is true of all the senses, of course, including touch.

But we are even further distanced from objectivity because even after our eyes have ''seen'' the view they would show us, we continue the selective process. Looking at the identical scene, you see a pretty girl and I see the clock which announces I am late. We have each pre-decided what we will and will not see because we want to use the scene to suit our different purposes. But even then we are not through with the various layers of subjectivity. All incoming sensual data is given to the mind for its demanding review. The mind then colors the data according to its own biases so that the data can become meaningful *for it*. Therefore, had I seen the girl, I would not have been charmed at all, while you continue to stand there with your mouth agape. The upshot of this is that the more we attempt to find answers, the more tightly we become involved in subjective processes.

There are also tangential considerations the scientific community itself is raising. What part does the observer himself play in what he observes? Does he perhaps even create what he observes? As a healer, knowing how easy it is to command an outcome for an obliging ''subject,'' I cannot help but wonder how many laboratory mice are unconsciously ''willed'' to be sick and/or well.

Unfortunately for the scientific procedure as it is presently mandated, nature has ordained that our experience of life be subjective. Scientists did not recognize this truth when they set up their methodology for arriving at facts, but they certainly should now because they are no longer naive. While they hoped to eliminate subjectivity from the scientific process, all they did was gerrymander reality to their own prescription and *call* it objective.

Moreover, their view of reality was constricted to fit a laboratory; reality was defined by the instruments on hand and the tests those instruments might suggest. All tests ever do, incidentally, is ''prove'' what the particular test is designed to ''prove'' (another subjective embarrassment). Therefore the results describe the test, not what is being tested.

A curious example of the way scientific methodology works is evidenced in the ''scientific'' investigation of psychic phenomena. The scientists' enthusiasm runs primarily towards demonstrating that a hoax is in progress. But if the means to this discovery eludes them, they proceed with careful measurements of the pulse rate or change in brain waves as if—by finding a feather—they could understand flight.

One scientist, with a lovely sense of the absurd, was asked why science was not more interested in looking into psychic phenomena with more seriousness. He replied that it presented problems to scientists and that perhaps scientists should be likened to the drunk who lost his keys one night. Someone noticed the drunk searching for them beneath a street light and when he was asked if this was where the keys had been lost, the drunk replied, ''No, but this is where it's easiest to look!'' This block to discovery is exactly what befalls the scientist when he allows his methodology to mark the line between what is acceptable (scientific) and unacceptable (unscientific) in the way of phenomena.

Science, in its wisdom, has decided that reality is physical reality, external reality—the visible and tangible reality of things or matter. Everything which is not ''real,'' according to this definition, is not considered pertinent and therefore exorcized from the research process. Included in this exiled segment are the sensibilities, the ego, intuition, the soul, intrinsic nature, the abstract, God (by any definition), luck, the mind, inspiration, creativity, motivation, the emotions, etc. All those aspects of life which give life meaning. They have, in short, eliminated the *nature* of life itself.

Thus it is that science can tell us the vitamin content and cellular makeup of an apple. They can reduce the apple's color to light waves and tell us why it falls, but they cannot discuss its essence—its apple-ness. Yet ''apple-ness'' is the only way we experience an apple. Nature is not physical reality, but physical reality as it makes itself known through inner, subjective reality. Scientists snip off causal reality and confine themselves to manifested reality. It is like studying the reproductive process after first deciding that there is only one sex.

Instead of attempting to synthesize the two recognized parts of the integral whole we call nature—the physical and the subjective—they opt for the analysis of one of these parts. Their way is division. With only one path in front of them (and perhaps the guilt of that division behind them) they have had to make an original wrong into a right by continuing the process of division until now we have sub-atomic particles. And still they give us no clue as to what nature *is*.

Their path may bring us more and more facts, but it takes us further and further from the truth. Truth must account for the sum of nature, not an arbitrary and artificial portion of it. It is time, I believe, to question the wisdom which has put us on this detour of misunderstanding.

*　*　*

The atrophy of human importance at the hands of science began innocently enough with Copernicus. Before Copernicus, we lived in a pretty cozy little world—a world designed with us in mind. The sun, indeed the heaven, revolved around us, supplying our needs of warmth and light. But hardly had we adjusted our disappointed egos to being part of a helio-centric solar system than Darwin proposed that we were the by-products of an evolutionary process which began, not in Eden, but in slime. Again we were forced to discount our importance.

Next we discovered that our galaxy is only one of billions. Again our significance shrunk. Now we have experienced nuclear destruction and its potential for obliteration. Aside from the fact that the nuclear industry seems to have a life of its own so that it grows exponentially like The Blob, the existence of humanity may well hang on a computer malfunction. It seems patently clear to me that humans have lost *all* importance in the minds of these latter day high priests and it's all because humans can't be tested in a test tube. It is all because we humans present ''problems'' to scientists.

On a tiny speck of cosmic dust, our very selves rooted in accident, we deal with a dragon who can destroy mankind in minutes. Is it any wonder we feel insignificant, impotent? Is there any doubt that if truth lies in some other direction we must find it?

* * *

As scientific answers began to accumulate, the aggregate knowledge started a process of de-mystification and a new form of natural law. God, to whom all things had been attributed, lost much of the power we had originally assigned to him. It was not God who caused the sun to travel the heavens as a gift to his children; instead the earth spun mindlessly on its course, compelled by the sun. It was not God who sent us rich soil from which we could farm our food; the soil was the residual dust from molten rock which had cooled and cracked. It was not even God who sent pestilence to cleanse the earth of the wicked; it was bacteria cultured under septic conditions.

You can see, can't you, why the Church put up such a bitter struggle against the challenges of science? Its hegemony was being threatened. Science, however, had made itself rationally irrefutable. Science dealt with facts, not faith. Besides, it offered something concrete, demonstrable, and *useful*. And it was a distinct relief to be able to build our beliefs on solid ground instead of having to grope down that long dark tunnel of faith with no sure end in sight.

Naturally as God's legerdemain was stripped of its magic and reduced to gravity, barometric pressure zones, and viruses, his ratings fell—and so did Church attendance. By the second half of the twentieth century, some people found the internalized religions of the East more in keeping with their spiritual needs than was Christianity. Some turned to cults which offered sanctuary to the spiritually lost in exchange for allegiance. Most people, however, were content to turn Sunday into a chance to sleep late and watch football.

While there has been an attempt to reintroduce God as love—in the best of marketing traditions—the compelling majesty of the old God is missing. Even among the devout he has become more of an amiable Santa Claus, ever vulnerable to his final collapse in the cauldron of knowledge. But the smart set has gone one step further. They have decided that a belief in God is a fine example of superstitious gullibility. In the 60's, ''God is alive and well and living in Argentina!'' was the cry of the stubborn theist against a sea of atheism.

I place fault for the eclipse of God's image squarely on the shoulders of religion. Had the Church pursued knowledge with the same vigor that science has, it might have had more to offer the willing faithful. Even though science has been eroding God's base like a robot run amok, this incursion has brought nothing new to institutionalized religion. Truth, to remain viable, must change, but the priests of all principal religions have written their tablets in stone, believing that wisdom, once found, remains unalterable and eternal. The result is that their message becomes less and less germane.

Religious leaders the world over have placed their emphasis on the deification of their specific founders, the sanctification of their particular paths to righteousness, and the enforcement of their particular moral laws. Beneath the panoply of ritual, the content grows sterile. The spiritual vacuum thus created cries for some credo—any credo—even a credo founded in the materialistic wisdom of science.

We ask so little of our leaders (and of course we get what we ask for). We wish only that they not give blatant offense in their dishonesties and that they offer us a small but reasonable hope that we are on the right path. We switched allegiance from religion to science, not because we had a choice, but because *we had no choice.* In the light of contemporary knowledge, institutionalized religion no longer offers plausibility. And so the voice of the scientist has become the voice of the new priest. His philosophical persuasions have become our doctrine, his material discoveries have become our articles (!) of faith, and his truths (for they are now truths, not just facts) have become our truths. The nuclear physicist and the medical doctor have become our gods.

Science promised, and in many ways delivered, the good life — something institutional religion had noticeably failed to produce. And science offered another bonus. Science made no demands. We neither had to mend our waywardness nor wait upon our reward. We needed only to take a pill or buy an appliance and put it all on a credit card. *This* god promised immediate salvation from suffering. Science was the antithesis of the religious lay-away plan of pay-now-enjoy-later. It was heady stuff.

But when we decided to put our faith in science and adopt its point of view, we cut deep into two support systems. One was the vertical family of supra-human personalities (God, angels, guides, saints, etc.) in which we had found such great comfort. We believed this hierarchy to be acutely involved with us personally, offering guidance, protection, wisdom, etc. Science substituted a horizontal cosmos of impersonal galaxies which offered nothing but vastness.

Before the emergence of science-as-religion, we directed our prayers heavenward, serene in the knowledge that God was listening and answering. Now we travel ''out there'' in the scientific version of a mystical experience — an experience which may be a tribute to our technology but does little to comfort our souls.

The second support system we surrendered was knowingness or intuition. And perhaps the grand champion of self-ignorance was Sigmund Freud. The fact that we are indebted to him for making the mind a respectable field of inquiry doesn't mean we haven't paid a price. Freud said that we did not know and could not know ourselves unless we were rich enough and adventuresome enough to seek professional help. And with this help what we find is very reminiscent of The Pride of Science—slime, composed of dark ugly motives, incestuous compulsions, killer urges and pathological fears. Yet, said Freud, it was within this sinkhole that everything relevant to human behavior transpired. Once again a wedge was driven between ourselves and our capacity to take care of ourselves. We perceived consciousness to be divided between a small surface awareness and a large subterranean, and largely unknowable, catacomb. Besides, who would want to know what lay beneath the surface? What was there was ugly—a brew of guilts and craven passions. And so . . . our self-image was forced to take yet another nose dive. (Does it strike you, as it does me, that history is one long sermon on how rotten we are?)

*   *   *

The field of psychology is a fine example of what happens to creative insight when it gets tangled up with scientific methodology. Psychologists first had to figure out how to make an overtly subjective subject "objective." They did this by designating the rat—what*ever* were they thinking of?—a suitable laboratory prototype for a human. A human, of course, was not sufficiently objective to be a proper subject for any study of himself. Take away the essence and study only what is left over. That's Scientific Methodology. Thus the glory of man was reduced to mazes, stop-watches, graphs and statistics so that round pegs could be put into the round holes carefully designed by scientists. Yet, the essence of man's mind is its creative *spontaneity,* its transcendent possibility, its subjectivity. What insight into its workings can be achieved with a stop-watch and a mouse?

The psychologist has become our spiritual leader. He is the agnostic's answer to all that is unreasonable in The Age of High-Tech. He remains fundamentally a scientist, however. Furthermore, his purpose is to offer immediate, not spiritual hope. His truths confine themselves to the ego, the libido, fears and guilts.

He offers no rich meaning to life, no larger context in which to take comfort. His prescriptions for a state of grace address themselves to nothing larger than social adjustment to sibling rivalry and sexual anxiety.

With our intuition debunked, our motives suspect, and our uniqueness ignored, it is hard to find anything about ourselves to admire or trust in the contemporary view. Indeed, our ability to make sound judgments is so challenged that we are expected to seek outside advice from ''scientifically qualified'' professionals on all matters. Do you believe, for instance, that it is advisable to have a yearly medical check-up? Then you agree with the scientist-doctor that you yourself can no longer tell if you are sick or well. The psychologist knows, the nutritionist knows, the surgeon knows, the meteorologist knows, the pharmacist knows—and it has come to this—even Dear Abby and the psychic know. But we do not know. Or so the experts would have us believe.

The fact is that each of us has perfectly good internal machinery which can always offer good guidance. To be told we cannot know and trust ourselves is simply terrible. I can think of nothing worse than this. If we believe it, it makes us impotent. Defenseless. It forces us to sell our souls for the security of first-strike weaponry. Such beliefs in personal impotence have also created a vacuum which has permitted the scientists to take over unchallenged. Without restraint or perceivable conscience we are now at the mercy of their chemicals and their warheads.

In part this state of affairs has come about because science has escaped our expertise. We have no way of judging scientific truths for we no longer really know what science is talking about, how it performs its magic, or how it arrives at its conclusions. We are helpless. This is the sort of thing I mean. Almost weekly, The American Cancer Society announces that another carcinogen has been found in a common household staple. While the Society makes these announcements in the interest of self-promotion and funding, what we, the recipients of such news wish to know is, is it credible? If it is, how come the scientists who invented the offending chemicals think otherwise? And why did they market it in the first place? Who do we believe? Certainly not ourselves for the controversy lies far beyond our expertise. And what do we do about it either way? Pray to a non-existent God called Love about our saccharin problem?

Indeed, the Cancer Society is so impressed with its little corner of the world that it very recently announced that if we were otherwise immortal, we would all die of cancer anyway. (Such a cheerful lot, these cancer buffs!) While such scare tactics certainly insure funding, it leaves an ungodly and unjustified fear in the hearts of the lay population. First, it is a patent lie. Not only is such a statement untrue, but it can in no way be demonstrated. It is nothing but the self-aggrandizing hope of dedicated cancer enthusiasts. But we are not amused. We are anguished—just as they wish us to be, for they are anguished too. But sadly, we are unable to pit our ignorance against their self-described enlightenment. And they know it. And it makes them unmanageably self-righteous.

When these kinds of proclamations are made, the casual reader must accept the headline. Once in awhile, however, there is accompanying fine print to these releases. It was, for instance, announced recently that the chemical solvent used to decaffeinate coffee was a carcinogen. Why did they think so? Because they had injected mice (?) with amounts of this stuff equal to as much as 12–24 *million* cups of coffee a day! Now I ask you! How many times have you drunk that much coffee in any given day? Whoever made that announcement was an unprincipled self-serving cheat. Take the calcium from 24 million cups of mother's milk and stuff it into a little mouse and you would produce a tiny stone monument. Then the American Calcium Deposit Society could go forth and fuel a fear about mother's milk.

Science, if it is to make itself useful, must re-evaluate some of the biases it brings to the very process of discovery. It must concede that the real world is subjective and it must adapt to that fact rather than force the facts to suit its wishes. It must extend science to its true meaning—the pursuit of knowledge, not simply the pursuit of what can be analyzed in a laboratory. We are in desperate need of knowledge right now, but we need the right kind of knowledge.

By and large, scientists prefer to live in ivory towers, isolated from ethics and a sense of priority. Most of them take comfort in numbers and find us troublesome. If you doubt this, you have only to remember that we live in the Nuclear Age—The Ultimate Age of Science—when 13,000 deaths from a nuclear reactor accident is a *hoped for* safety goal. How characteristic it is of scientific thinking to plan city evacuation procedures so that we have nothing to fear from nuclear war. And how well those plans fit together mathematically. So many cars moving at so many miles per hour can evacuate a city in so many hours, they say. We try to ''evacuate'' cities daily around our population centers and this usual rush-hour is not complicated by panic, unpreparedness and the simultaneous departure of all its citizens. We also rely on the Highway Patrol, ambulances, towing companies, gas stations, etc., to get things untangled. Still it is a mess. Besides, we all have some place to go when it is all over! What happens when 2,000,000 people begin to mill around in the suburbs?

We live in the midst of radioactive wastelands, Love Canals, and Three Mile Islands—all the by-products of an uncaring and irresponsible scientific community. Yet the avant-garde of this community is searching for a quark, a theoretical entity. We go to the moon and beyond because it lies within the scope of mathematical niceness. But 28 of us die every single minute because solutions to famine do not submit themselves to comfortable equations. Nuclear proliferation is an exciting challenge but the knowledge needed to effect detente does not lie within the perimeters where science chooses to look. Imagine what could be accomplished if the same amount of money, time and *enthusiasm* which we now spend on trying to destroy life were spent investigating the mechanisms which would insure life.

*   *   *

In terms of our cosmological origins, The Truth which science has given us rests almost exclusively on two theories. The first is Darwin's Theory of Evolution and the second is the Big Bang Theory of Creation. Teachings and theories are strange, however. They are accepted or rejected, not because they are true or false (though that is always the contemporary opinion) but because they confirm or deny a direction which has already been established. Christianity gave new meaning to authoritarianism. Darwin offered a rationale for breaking away from the stranglehold of a simplistic theology. Einstein and nuclear physics was the logical direction of dissection.

When a theory proves attractive for whatever reason a very subjective process of winnowing the raw material begins. The data which fits the theory is noticed and that which doesn't fit is either ignored or reinterpreted. Pretty soon the theory begins to fit more and more remarkably until at last it is indisputable. A theory, then, becomes self-substantiating if that is the wish. It is important to look at both Darwin's Theory of Evolution and the Big Bang Theory—these explanations of ourselves and our origins—because their implications so degrade the human spirit and condition. I have no objections whatsoever to being the offspring of an ape. That is certainly not debasing. What is so belittling is that we have been reduced to nothing but a biological system. There is not the slightest acknowledgement of our unique subjective qualities. In metaphor, science has reduced the *Mona Lisa* to nothing more than canvas and paint when they falsely state that we are an accident of energy and aggression.

In addition, all ties to a larger overview of existence have been severed. From the point of view of scientific truth, nothing exists but the physical world scientists study. Put in religious terms, they have eliminated God. The result is that those who believe what science has to say, must deal with isolation and cosmic indifference. Recognized or not, cosmic indifference is as difficult to face for an adult as abandonment is for a child. Babies curl up and die if they are not fondled and cherished. They *must* have connection . . . and so must adults.

Too many of us have been deprived of this connection because our new religion (science) does not admit its possibility. Feeling this lack of union, we concentrate on sex and ''meaningful relationships'' as the context of life collapses into the penultimate circle of two. Two is not enough. Besides, such intra-reliance forces a relationship to carry burdens of mutual support it cannot possibly sustain.

Integrity is defined as ''the condition of having no part or element wanting.'' Science, in discussing us as *only* physical beings, deprives us of our integrity, for absent from their description of us is everything having to do with our inner selves—our emotions, our minds, our souls, etc. And therefore their truths lack the same integrity because parts are missing.

*   *   *

Most knowledgeable people accept as fact Darwin's Theory of Evolution, its principle of the survival of the fittest, and its implication that might makes right. If they also believe in God, they find God's way to be the way of evolution. He didn't, they would say, produce Adam and Eve; he produced viable slime—which is almost as magical, I suppose. Viable slime is certainly something science can't produce.

Darwin's Theory has had its deep and disturbing effect upon society for it has both explained (?) and justified the killer instinct in our political and economic life. We have problems enough with our baser promptings without being told that there is a biological rationale—indeed, a biological responsibility—to act upon these urges. More cutting, however, is the Darwinian Answer to the big question Why? We are, or so we are told, the chromosomal result of mutations. Nature/God doesn't give a damn except as we biologically improve the chances of the species. It therefore means that those of us who are able to produce children who can adapt to a polluted environment will endure in our progeny; otherwise, our presence is *de trop*.

The Darwin story is that once upon a time a Naked Ape thoughtfully picked up a stone with the intent of committing mayhem in the name of survival. The results were so outrageously satisfying that soon apes who threw stones became humans. (Why we still have an ape population is not explained. Shouldn't they have all died off or gone on to better things?) Sensitive to human perversity, I have little doubt this first stone-throwing ape felled his neighbor, but the story is vague on this point. From that moment on, the story continues, we have been learning, ever learning, these exciting survival skills until triumphantly we are at long last able to destroy *all* our neighbors as well as ourselves in the service of this marvelous mechanism which insures *survival!*

If, in viewing history, we think we are witnessing the evolution of our species, it is because we are, more accurately, looking at the evolution of our belief in our abilities. For example, until a few years ago it was thought impossible to run a mile in four minutes. Once that "impossible" barrier had been broken, however, it became "normal" to run that fast. Now all world class runners must do at least that well. It is not our bodies which are intrinsically improving so much as it is a shift in our mind-set. We have raised the false limitations we had set upon our bodies. It is man's self-image which is undergoing evolution, not his innate capacities.

Belief in Darwin's theory has set us back in discovering what a human truly is. In concentrating upon our supposed bestial origins we have overlooked our differences. In part, I suspect this stall on the path of knowledge has been greatly exascerbated by Darwinists who refuse to give up. In the 130 years since this theory was first put forth, not one shred of empirical evidence has shown up to support Darwin. No species has ever been found in the process of transition.

Scientists may argue that 100 years is nothing as evolutionary time is measured; but nonetheless, under Darwin's theory, something should be continually afoot, and it is not. The apes we know about get along very nicely without throwing stones. Yet, if Darwin is correct, then every so often an ape should produce a humanoid offspring and this, very noticeably, is not happening. Apes go right on producing little apes, no ''better'' than their parents.

There is, within Darwinism, an arrogant assumption that some life is better than other life—that the life of a laboratory mouse, for instance, is less important (it being less ''evolved'') than human life. To *whom* it is less important, of course, is left unanswered. Hitler's visionary attempts to create a Super Race is nothing more than the logical extension of this kind of thinking. If nature does this, why not man? Couldn't men do it faster, better, and more discriminatingly? If there is a difference between the best of life and the lowest of life, then there is also a difference between the best of ethnic groups and the lowest of ethnic groups. Many racists see nothing wrong with this assumption, but many scientists do not understand that they share the same kind of bigotry.

The pictures and displays we see of the evolutionary process have been designed to fit the scientist's imagination and hopes. It is as if we took a dozen leaves from a dozen trees and arranged them by size in order to prove that leaves are getting larger.

Scientists look at the fossilized skeletons of A and B. Everything looks pretty much the same except that fossil B has, let us say, little paws. ''Ah-ha,'' they say, ''A must have grown little paws (assumption 1) so that he could walk on land (assumption 2) in order to find more food (assumption 3)." I find it interesting that this kind of forcing is not attempted on the vegetable kingdom. At least to my knowledge, no one has bothered to trace the rose back to some deep sea algae. Apparently we are able to accept diversity for its own sake in the plant world.

Darwin's theory implies a narrowing process, for obviously, if only the best survive, there is only one best way to fly, one best way to swim, one best way to catch insects in a web—all of which is demonstrably not the case. There are, for instance, some 8,500 kinds of birds—all perfectly adapted to the art of flying. According to Darwinists, all those birds came up a hierarchical ladder from single celled life, studying biologically how to be the very best sort of bird. Yet, to date, there appear to be 8,500 "best" ways to do that. Frequently I hear scientists exclaim delightedly upon this diversity and then turn right around and offer it as proof of nature's marvelous process of evolution—a process which supports singularity.

If any one fact suggests that nature demands diversity and does not function on the principle of selectivity, it is those 8,500 species of birds. The fact that animals are so beautifully equipped to survive in a myriad of special ways suggests to me that within the laws of nature almost anything will work, including the hippopotamus. It suggests a cosmic thrill in seeing how many designs *can* work to perfection. It does not suggest an attempt to find the Ultimate Design.

It seems highly likely that the creation of a particular species simultaneously creates a benign environment for that species. That is, the environment is not plunked down like a garden and then overlaid with animal life which lives or dies in its proffered or denied hospitality. Instead monkeys, let's say, who like trees have as much to do with the creation of those trees as the trees have to do with the presence of monkeys. In other words, environments are a unified *one* in their flora and fauna—each cause and effect of the other. There is sublime cooperation here, not competition.

Darwinists do not mention de-volution. Who is to say that fossil B didn't lose its little paws to become fossil A and breathlessly return to the sea? If the process of evolution is real so must be the process of de-volution. They are integral partners. And if this is the case, perhaps apes are not our forebearers but an example of what happens to little boys and girls who do not eat their vegetables.

Aside from the arguments mentioned above, there is no evidence to support the notion that Mr. Strong's genes *do* produce "better" progeny. Here is another assumption. It is certainly not true in the human species—to the bitter consternation of many a father and son. This includes the elder and junior Darwins specifically. Charles was a distinct disappointment to his father because Charles did not wish to become a minister. But among his so-called failings he was also weak and sickly—a biologic atavism. Under these circumstances I have to wonder how much of Darwin's theories were colored by his own sense of inadequacy. We do know that aggression grieved him; he was appalled by its omnipresence. Did his theories perhaps reflect a bitterness for "the way of nature"?

All these observations are petty, however, in comparison with the basic flaw in Darwin's theory. *Survival depends not on competition but on cooperation.* No pup or chick could survive unless the instincts of concern were not built into every species. The very act of procreation itself is an act of cooperation. Even the so-called "territorial imperative" depends upon the cooperation of other animals to abide by the unspoken rules. Pecking orders are also systems of cooperation; the leaders are permitted to lead, and the others are content to follow. Anarchy is the antithesis of cooperation and anarchy doesn't exist in the animal world unless you include humans, and even we run from it.

What *does* happen in the animal kingdom is that carnivorous animals need each other to exist at all. They must eat one another. The ''strong'' of one species can survive *only* because of the ''weak'' of another species. Paradoxically, it is the ''weak'' therefore who actually effect the survival of the animal kingdom. Existence itself is therefore owed to the weak, not the strong, for the fox cannot exist without the hare. That is quintessential cooperation.

It is also clear that there is a qualitative difference between being human and being anything else. If we are descended from apes, as science would have it, how does this account for some very remarkable qualities about *our* species? From whence came our imagination? Our love of music? Our capacity for laughter and tears? Our obsession with revenge? Our ability to think abstractly? Where do we find in the ape the origins of our creativity, our talent for language? There is no real sign of any of these endowments as we move up (?) the biological ladder in the way we think we see forerunners of arms and legs in quadrupeds.

The distance between apes and humans is enormous. It is a chasm. Darwin himself was quite aware of this flaw in his theory. He suggested a Missing Link. When discovered (a very active search has not done so), the Missing Link was expected to simultaneously display the brutish characteristics of an ape and the incipient talents of a Picasso. . . . Well, on second thought, I think Mr. Link married a friend of mine and I may still have their address if someone wishes to pursue this matter. But unless this man satisfies the needs of Darwinists, the theory, in the final analysis, rests its case on something which is *missing* from the available data. I'm sure any lawyer would be grateful for the permission scientists have granted themselves.

*   *   *

I cannot find one particle of difference between the Creationists' explanation of creation and the scientists'. The authors of

Genesis looked around them and saw what composed the world—earth, sky, plants, animals and humans. Working backward, they conceived an explanation to explain this end result.

The universe of a scientist is not made up of these obvious things. For him, it is composed of energy, matter, and antimatter. (To a scientist, people are chemicals or matter.) In other words each theorist—the creationist and the scientist—took data familiar to him and imagined the kind of event which would produce his kind of world. The effect was mother to the cause in both cases. It just *sounds* as if there were a difference because the "things" of each world are so differently described.

Let's look then at the Big Bang Theory of Creation which would deprive us of our spiritual inheritance and see if it has or has not advanced the cause of knowledge.

The Universe, say scientists, *was created from a naked singularity.*

And what may I ask was Adam? "Naked singularity" is a phrase invented to cover an event which can only happen once, although such a one-time occurrence is not supposed to be scientifically permissible. But then I like to think we are all permitted a little back-sliding now and then. The scientific view rests on the assumption that the universe is expanding, presumably from a central point—a theory I challenge in *The Theory of Laminated Spacetime.* The Big Bang thesis incorporates several theories about black holes, Planck mass, and the like. All of it has more to do with creative mathematics than it has to do with pragmatics. Stripped of its cult-obscurities, it simply means, "Once upon a time . . ."

This singularity exploded in a nuclear reaction. *

We are not encouraged to ask why. I find my own imagination takes more easily to an industrious God than to a spontaneous explosion. It still means exactly the same thing only the offending

---

*A nuclear explosion either through fusion or fission is, I believe, antithetical to nature and cannot happen anywhere "naturally" (including the sun). Physicists are not getting energy *from matter* in nuclear destruction as they presently believe. They are punching a hole in the time/space webbing of physical existence. See *The Theory of Laminated Spacetime.*

word *God* has been expurgated and a scientifically acceptable event substituted.

The explosion released energy, matter, and antimatter.

These are the constituent parts which science needs to find in such a one time event because these are the component parts of their universe. The statement, ''God created heaven and earth,'' has every bit as much validity. Scientists don't like vague descriptive words like heaven and earth; they prefer vague descriptive words like energy and matter.

When the explosion cooled most of the matter was destroyed by antimatter but some of the matter was left over.

Scientists believe that when matter is created, antimatter is also created which destroys matter. With this theory there should be no such thing as matter, but this doesn't work out very well. The loophole which allows escape is the Principle of Uncertainty—a sort of physicist's Murphy's Law—which says you can't be sure about matter. Such an explanation has no validity in reason; it is a cover-up for ignorance. In the reverse of this, Eve's emergence from Adam's rib cage is also a cover-up—a cover-up for knowledge. Carnal knowledge.

The particles formed hydrogen and helium.

How???

Hydrogen and helium are the building blocks scientists need to explain their universe. They do not need people. Otherwise they would have called these particles Adam and Eve as did a former generation.

These particles, shooting out from the naked singularity, formed galaxies and stars.

Just like that! At this point we have moved from total chaos to total order with no explanation whatsoever. It is a leap of faith which beggars the imagination. Yet, science rests its reputation on the certifiable. Leaps of faith are not permitted.

Why didn't *this* explosion behave like all other known explosions and make so much pixie dust?

And do you not find it interesting that both the Creationist and the physicist rely on *dust* as the physical source of all?

If you believe the Big Bang Theory—and you probably do—it is because you have come to trust science in a field you can't untangle for yourself because it is explained in tongues with an air of certitude. What science has done is recast the *Book of Genesis* using scientific terminology instead of something we can understand. Nothing has changed in terms of de-mystification, in terms of logic, or in terms of fact. Indeed, only one change has been made; man as a meaningful ingredient in the origins of the universe has been eliminated from the scientific version. Science does not want to deal with man except at his cellular level. Man defies scientific methodology. So science has eliminated him from its theories and truths. If you like the Big Bang Theory, you might as well go back to the Creationist's point of view. Then, at least, you can put yourself back in the picture.

For a scientist, the universe is exciting because it is a mathematical marvel. It is a machine of superb design and mystery. It is marred only by man's presence, which does not submit to mathematics. What the scientist fails to understand is that the universe has been designed for a purpose (which also has no mathematical equation). That purpose is the intimate experience of selfhood for all of nature's many manifestations. The universe is designed for the sun as much as for the ant, for the spruce tree as much as for the human.

5.

The Theory of Laminated Spacetime

W E HAVE BEEN DISCUSSING the traditional truths concerning our origins and purposes which most of us take to be true in great part. I certainly hope I have left no doubt as to what I think of these truths. I would like to throw them all out except for the intuited knowledge that there is something extant which is greater and more powerful than we humans. For the moment we can call that something God but we will certainly have to redefine that term as we move along in our discussions.

I would like to exchange these traditional, and not very helpful, beliefs for a completely different understanding of the universe. I call this new mock-up the Creating Cosmos, believing that the system upon which All That Is[1] is founded, is the creative act itself. The cosmos is not a thing—a noun—it is a verb, a perpetual activity. All That Is is both sponsored by and the cause of activity. And All That Is exists within the creative acts of itself. If this is not sparkling clear, it will become so during the course of our ensuing explanations. Before we can detail the Creating Cosmos, however, let's look at the suggestion my father made to me about the constituent property of time and space which would not only make the creative act possible, it would make it mandatory.

[1]A phrase borrowed from Jane Roberts' *The Unknown Reality: A Seth Book,* Vol. 2. Prentice-Hall, Inc., 1979.

My father said that it had occurred to him that spacetime (physical reality) might be blinking off and on so fast that we could not detect it. That is, he explained, a condition where everything existed would be followed by a condition where nothing existed (a blank) followed by a condition where everything existed once more. Because of the great speed of these events (I have since put this at the speed of light), physical reality would *appear* to be a continuum while actually being an intermittently broken sequence of spacetime laminae in the same way that a movie film expresses motion even though it is composed of a series of still pictures interspersed with blank bars.

Of course, that blank—as he expressed it—wasn't a blank at all, but the spaceless, timeless creating medium Holmes called Universal Mind and religions have usually described as God, the source of all creation. Why? Because when space and time stop, we have, by definition, a condition without either space or time. And because "the blank" itself must have the capacity to produce the next lamina of spacetime.

Those of us who feel comfortable with our intuitive responses—who say we *know* though cannot prove—are always abashed when others are skeptical of what we feel so surely to be true. I cannot take you further towards an acceptance of this theory than to say it is true for me. Nor can that blank space which is without dimension ever be measured; it can only be inferred. To help my case along, however, it wouldn't hurt to remind you that down through the ages the very wisest among us have *unanimously* counselled of an invisible and intangible "soul" world. They have *unanimously* believed that this invisible world was the source of all creation. They have *unanimously* agreed that the proper life direction was attunement with the life force offered up by this world. All that has been missing over the centuries is the mechanism by which the two worlds—the world of spirit and the world we can see—and measure!—might fit together.

The universality of such intuited knowledge forces a conclusion which must be faced. Either all probing minds across all cultures and all centuries have been deluded into reaching the exact same conclusions or any theory which purports to offer a cosmological scheme of how the universe works *must* incorporate such time proven wisdom into its thesis. To date, scientists have been unwilling to accept this responsibility. It's as if they were attempting to explain cookery without mentioning food, or fire without mentioning heat.

When I began my study of physics, I repeatedly found statements like, "This behavior could only be explained if the particle in question actually disappeared," or "It's as if all motion stopped only to be resumed again." But such a thought always died a-borning

What I like so much about this theory is that it quite literally cracks open physical reality to force a blank in the scheme of things for a legitimate combination of the world of matter and the world of subjectivity. And it does so on science's own terms. This is not the philosophic mind speaking of its intuited knowledge. *This is a physical mechanism.*

If it is a fact that spacetime is laminated, then science must eventually accept this fact. If it is true that all which exists dissolves into nothingness only to be created from that nothingness, then this nothingness will have to be taken into serious account. If this nothingness has the power to create and recreate billions and billions of galaxies in an infinite expanse of time, this Source of All can no longer be ignored by science. And that will be a bright day for all of us.

Scientific theory already embraces the concept of black holes. Scientists are therefore already prepared for the theoretical disappearance of matter. What they may not be prepared for is the emergence of Causal Source, or what is more commonly called God.

Laminated Spacetime is a description of the creative process. It is the mechanism which permits the existence of all physical form. It is the Prime Act. Using the Theory of Laminated Spacetime, we can detail that act of creation.

I never did ask my father what gave rise to his theory. Long before we discussed it, he reported that he had had a dream vision—one of those once-in-a-lifetime events which comes as an answer to a question long asked. He was told, he said, that the entire universe was built on the single principle of creation and destruction. Perhaps his theory arose from this, or perhaps, emboldened by such information, he went on to experiment intellectually with its implications. In any event, the process of laminating spacetime demonstrates a truth which comes in the form of a paradox; it is only through the destruction of what has gone before that creation can take place.

Destruction is an integral and essential concomitant of the creative process. The egg must be broken to have a chick. The blank canvas must be obliterated to create a painting. Blessed singleness must be destroyed to contract for marriage. The act of creation itself is founded on destruction, and there can be no more basic statement of this truth than the continual destruction of what has just been created in the process I call the lamination of spacetime.

*   *   *

Before we look more closely at what Laminated Spacetime infers in terms of spiritual intent, I would like to introduce two new words. The first—a synonym for the blank of no-spacetime between two physical laminae—is *endosphere,* the reality within. The endosphere does not blink off and on. It is, by definition, without the dimensions of space and time. It is therefore without beginning and without end, both spatially and temporally. The concepts of beginning and ending can only apply within the dimensions of time and space. The endosphere is eternal and infinite. It is everywhere and nowhere. It is the scientist's geometric dot denoting a place of non-existence which nonetheless must be described. It is ALL. Actually there is nothing but the endosphere yet under the temporary conditions imposed by spacetime it momentarily assumes a specific form. It becomes matter-in-motion. Note please that I did not say energy and matter, but matter-in-motion—one six dimensional unit—a statement which is more fully explained in the book on physics. For the moment, imagine the endosphere to be a ship's signalling lamp which is turned on for all time.

The second word is *ectoform*. An ectoform is anything which manifests physically. An ectoform can be a universe or an atom. It can be an oil painting or a buffalo. It can be a solar system or an infected finger. It can be a marriage or a union strike. What is considered to be a unit of matter-in-motion or an ectoform is in the mind of the describer. We are free, therefore, to make an ectoform as inclusive or exclusive as we wish. Cosmically, a discrete "thing" must be described by its coalescing activity and intent, not its physical boundaries. A cell has its boundaries but it also acts in concert with the body to which it belongs. Events are but the actions of bodies, acting in concert with other bodies, in larger gestalts. An ectoform, then, is the endosphere defined by the dimensions of space and time and given a particular classification by ourselves.

Imagine now that the dimensions of space and time work like a shutter which has been placed across the face of the lamp, the endosphere. As the slats close and open, the light blinks off and on. We see the light (the endosphere) only in its design and dots and dashes which out-picture ectoforms for our peculiarly and specifically organized senses. We "get the message" because the infinite endosphere has been made finite in a way we can appreciate. The endosphere is All That Is. I think of it as *the purpose, the power, the process,* and *the product.* But in order to simplify our discussion the term endosphere will hereafter include only the purpose and the power. The lamination of spacetime is the process, and an ectoform is the product.

The endosphere is creative potential; it is infinite possibility—the agent as well as the means of change. It is the power behind what a physicist usually describes as energy, when that power manifests physically. Energy is more accurately the action taken by the power of the endosphere. By itself, the endosphere would always remain an undefined, uncreated whole, like a potter's bucket of clay. It could be nothing more than unexpressed potential as if the clay yearned to express itself but could not do so. Only by limitation and definition can a *thing* be created—can an ectoform be brought into existence, can a clay pot take form.

But an ectoform by itself (which is not possible) would have nothing to express nor the power to form itself or act. By imposing the limits of time and space upon ALL, which intrinsically has no such limits, we arrive at a physical definition of ALL into some of its possibilities. The Theory of Laminated Spacetime describes the process by which unlimited potential and finite possibility are synthesized to produce physical reality. Physical reality, or the ectoform we call the universe, is an outlet for the compelling creative needs of the endosphere. *All ectoforms are the effect of the endosphere.* Thus true cause can never be found within an ectoform itself, but must be accounted for by the laws of the endosphere.

Another way of perceiving the blinks created by Laminated Spacetime is to assign to them the necessary task of forbidding the entry of too much power into our reality, a reality which is built to absorb this power in small incremental doses. Our environment and all within it are synchronized to this limitation. In nuclear destruction, the blinking mechanism has been damaged sufficiently to remove its safety features. The bonding ability of spacetime has been destroyed. Momentarily we get a whiff of the unchecked endosphere. This increase in "energy" is coming from the endosphere, not from the destroyed matter. (See *The Theory of Laminated Spacetime.*)

Although it is too early in our discussions to amplify on the creative process, suffice to say at the moment that the endosphere is the source of an ectoform in the same way that the activities of a television sound stage are the *cause* of the program. The program is the effect of the unseen effort which has already taken place on that sound stage. While there is truly no beginning and no ending in the greater reality of the endosphere, if you must think of a beginning, physical experience begins with a Big Blink. The action is already in progress as if you had just turned on the TV set. Under such circumstances, we are not treated first to a bare stage, then to a gradual decoration of that stage, until late in the show, half formed humanoids appear to mature into the characters of our little playlet.

Human beings have been around "forever." If we find no traces of other great civilizations it may be that we are looking in the wrong places—in deserts instead of the depths of the oceans, for instance. Or perhaps their remains are 10 miles below the earth's surface, thrown there when the earth folded in upon itself. Then again perhaps they existed on an earth which was 10 miles greater in its radius, and that larger earth has either been eroded or nuked to its present size. I admit to losing patience when the scientific community presents their findings to us as incontrovertible fact, as if their limited imaginations had, of necessity, to contain the whole truth. Why don't they at least say, "This is the best we can discover at the moment, but who knows?" Why don't they leave things open ended so that others will not be so intimidated to explore different possibilities?

Such beginnings as imagined by our present-day conceptualists may appeal to our present-day logic, but it is primitive logic, hamstrung by our concepts of time and space and their limitations. We need to leap such boundaries. The universe is ordered up from a timeless, spaceless condition which only momentarily presents itself as a finite and dimensionalized condition.

What we see on our "television screens" is a combination of the writer's intent, the producer's decision to make it "real," the director's sensitivity to the overall purpose of the proposed show, and the various performers' abilities to support the entire effort. In other words, what we see is the final statement of a great deal of creative activity which we do *not* see. And yet it is this unseen activity which is the cause of the story we watch unfolding.

We are so caught up in the story line of what we watch on TV that we suspend judgment and believe that boy gets girl because he is so charming. This is not true. Boy gets girl because the unseen writer chooses to materialize that possibility within the script. The writer makes the choice, not the boy or the girl. I am not, incidentally, building a case for God, the Unseen Dramatist, as if we were but the puppets of an unseen force. We write our own scripts, consciously recognized or not, but we do so from the unseen or endospheric/inner level. When we decide to fall in love, for instance, we do so from our inner dimensions. Only after that inner decision do we find our beloved charming, not before. If misunderstandings arise as to where cause lies for us who live in the picture all around us, it is because we have been conditioned by our own concentration and by the materialistic outlook of scientific thought.

I like to think of the relationship between cause and effect this way: all the interesting and necessary part is going on within the unseen endosphere but we are so literal minded, so in need of specifics, that we wouldn't believe it if we didn't see it. So an arrangement has been made which flashes a dimensional picture, or progress report, with each blink of Laminated Spacetime. The picture says, "This is what is being created by you and others . . . And now this is what is being created by you and others . . . And now this . . ." Parenthetically the picture is also saying, "If you don't like what's happening, create something else." Life is a creative process. Your creative process. What you experience is what *you* have chosen to create, whether you acknowledge it or not.

*   *   *

The endosphere is Infinite Power. When a finite amount of that power is made physical by defining or limiting the infinite in the process of lamination, it becomes finite matter-in-motion. Matter-in-motion displays itself simultaneously as both matter and motion. Matter and energy cannot be separated from each other as is commonly imagined. Nor are they two forms of one kind of thing. There is only one form of one thing — the endosphere made finite. Think of a tennis ball. Its ability to bounce is a property of itself. You cannot make it bounce more by battering it until its *form* is transformed into its activity as nuclear physicists claim.

The Theory of Laminated Spacetime states that total reality is a sequence in which time and space exist and in which they don't exist. If what we observe has dimension and exhibits some activity, then it is an ectoform. Remember that even something as apparently phlegmatic as a rock is attached to our planet and is therefore hurtling through space in a state of high activity. An ectoform is physical, momentarily set off from its larger constituent self by space and time. An ectoform is the *effect* space and time have upon its greater self or its endospheric aspect.

If what we intuit to be part of life cannot be located physically, then it is timeless and spaceless by definition. As such it is of the endosphere. Not only does this mean that it has no physical properties but it is cause — in some form — to physical form. Among those elements of our total selves which are of the endosphere we can count our souls, our minds, our consciousness, our emotions, our psyches, our personalities, etc. These are all different descriptions for the non-physical or endospheric self.

This non-physical self rests very securely in the timeless-spacelessness of the endosphere. It therefore endures infinitely and eternally. It is our bodies which are blinking off and on, not our souls. This is one reason why such a theory may be so hard to credit. The *feel* of reality is that it is, indeed, continuous, from the point of view of that which is doing the perceiving—the mind. We are aware of no disruption because, for the mind, or the self, there *is* no disruption.

Simply put, this makes all of us immortal. Laminated Space-time gives you, if you need it, "proof" that you have an immortal soul and cannot be extinguished by the experience of death. Each of us endures in a state of infinite eternity because each of us is of the endosphere. Only our bodies are finite. Furthermore, reincarnation is no more mysterious than our physical ability to be a teacher, a parent, and a tennis player "all at once." We can be all of these personalities simply by focusing our awareness on one project or another. In physical reality we have to perform the associated activities in sequence because that is the way of physical reality. But in the timeless condition of true reality, all incarnations are in the eternal now.

The passage of time from which our sense of history is derived is enculturated. To see this, imagine that no records were kept and that every reference to time and date carried the same designation. Suppose that throughout history all times and dates were 12:21 p.m., January 8, 2009. Pretty soon the sequential occurrence of events would begin to carry little weight. What would be important would be the emotional impact of what was experienced. Emotionally charged events would be "nearer" to us, more vivid for us, than the events which carried little or no emotional content. In the truer reality of the endosphere emotionally charged events feel as if they happened "yesterday." Unimportant events are "ancient history." Experience enters a pool where a sense of linear time has no meaning. The mind creates and re-members from the *eternal now*. It creates (recreates, if you will) the past and the future from this vantage point. Only our enculturated obsession to record makes history and perception of our passage of time "real."

So, too, are we indivisibly one with All That Is. In the conditions of the endosphere we are spatially and temporally without separation because we are without dimension. This has some very specific ramifications when it comes to the conscious use of mind as a creating tool, but for now just let's say that telepathy (the transmission of mental messages) takes place under timeless, spaceless conditions. This is why scientists have such great difficulty with the phenomenon. In their world such a condition is impossible. Yet this is sheer nonsense. The very concept of space suggests the possibility of no-space.

By now I hope you are a bit ahead of me. I hope you are wondering how nothing can be turned into something, for this is what we will take up in the next chapter.

6.

I-ams, Big and Small

MENTAL INTENT creates physical things and events or ecto-forms. Before we can really grasp what this means we must come to it much more slowly. For this purpose, I have chosen to take you on the same intellectual path I took. It may be a bit circuitous, but it's safer if I wish to bring you along with me.

All physical phenomena follow the same basic behavioral patterns whether these phenomena manifest as an electromagnetic wave or the birth and death of a solar system. Physical manifestations expand and contract, wax and wane, ebb and flow. All behavior vacillates between the workable extremes of creation and destruction as the vitality behind that behavior increases or diminishes. When the frequency between the crests of these activities is of short duration, we call it a vibratory frequency. When the span between the crests is longer, we call it cyclic behavior, or the seasons, or birth and death, or boom and bust, or any number of other synonyms. Some of these events are accomplished in a single blink of spacetime. Some take trillions upon trillions of blinks to complete one cycle.

My father's life's work was the study of many of these cycles—the rhythmic, predictable tides in the affairs of this planet which underlie all activity. From the price of cotton to the width of tree rings, from mine cave-ins to aluminum production, these accurate rhythms can be statistically demonstrated. What's more, they repeat themselves for as long as we have any records from which to extrapolate the data. What began for my father as a statistical oddity, had, by the end of his life and his unceasing labors, become an acknowledged and incontrovertible fact. Cyclic behavior is not only a natural phenomenon, but it touches all activity, including the very size of the earth which is demonstrated by the phenomenon we call the nutation of the earth's axial pole. (See *The Theory of Laminated Spacetime,* Chapter 11.)

What makes casual observation elusive is that many cycles are likely to be reflected in any one activity—some cycles going up and others going down so that the net result is apt to appear quite haphazard. For example, if we take a 4.89 year cycle (A) and a 5.5 year cycle (B) and a 6.07 year cycle (C)—all found in stock market activity—and we synthesize them, this is what we get: (See Fig. 1). Not very illuminating for the uninitiated, is it?

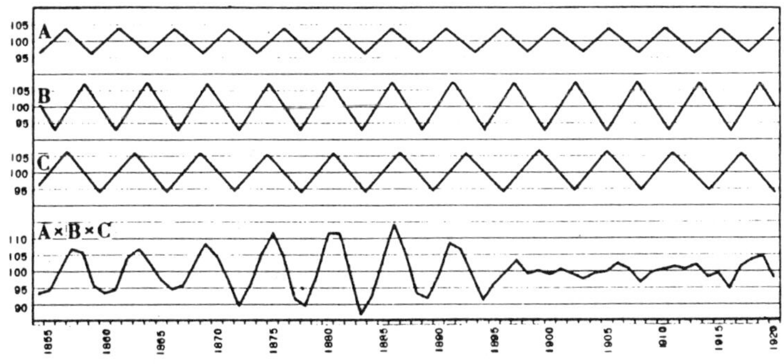

Fig. 1: 3 Cycles Synthesized

Used with the permission of the Foundation for the Study of Cycles.

But more interesting still than cyclic behavior within any given activity is the fact that a single and specific cycle runs through a broad spectrum of activities. For instance, there is a 5.9 year cycle which appears in business failures, copper mine production, grouse abundance, and sunspots. Incidentally, sometimes one of these cycles is mistakenly identified as the cause of another. Such a mistaken correlation has been made between smoking and lung cancer. Already the statistical evidence—and that is all there is, by the way—is beginning to fall apart in some areas.

The indisputable evidence of such cyclic behavior suggests to me that the universe is run on a system which demands that all of its creations change. In order to exist at all in a physical dimension, change must occur within that very existence. *To be* carries with it the necessity of *becoming* (something else). It is implicit in the dimension of time and our experience of the use of energy. The hours of daylight must change each day, the lungs must empty and fill, the flowers must bloom and die. We take for granted the rhythms of the seasons, birth and death, night and day, but there are just as predictable cycles in housing starts, locust populations, women's hemlines, and landslide elections. All must change and all change must occur in its precise rhythms according to laws we know almost nothing about.

I once suggested to my father that he assume that there were cycles of every wave length from nano seconds to the 220,000 year cycle found in solar activity and see what kind of profile this might present. But listening to his daughter was never one of Dad's strong points! Had he taken me up on my suggestion, I expect he would have discovered that there were an enormous number of resonances—call them cycles or seasons or electro-magnetic waves—and that these resonances clustered around some frequencies but were noticeably absent around others. In other words, there would be cyclic behavior going at "right angles" to the ones originally charted. These lateral peaks and troughs might reveal another sort of sensitivity profile. For example, there *might* be more kinship between whale populations and international conflicts than between whale populations and krill populations. (I said might! I am only inventing a correlation to illustrate a point.)

Our present ideas about biological classifications might have to be drastically overhauled. I don't think that would hurt a bit. As any ecologist can tell us already, there can be closer bonds between the tree by the river and the fish within that river than there are between that tree and another of its kind across the country.

Then, just to keep everything jumping, I would also expect that cyclic behavior itself was "moving" or undergoing change. A 5-year cycle, for instance, might be alternately moving towards becoming a 5.5-year cycle or diminishing to a 4.5-year cycle, as the cycle itself exhibited cyclic behavior.

Finally, the timing in one latitude differs from the next. (See *The Theory of Laminated Spacetime.*) This means that my father's figures will all have to be recast—according to the latitude in which the activity took place. Such figures may not be so easy to come by because the gatherers of such facts have lumped everything together as if there were only one clock in the universe. (No wonder life appears so chaotic. Its order is so subtle!)

*　*　*

The more I thought about cycles, the more it seemed to me that if matter-in-motion were to act in conformance with the laws which cyclic behavior described, then there had to be a mechanism within each ectoform, or physical manifestation, which "decided" how much of the infinite enabling power of the endosphere it was going to use for its purposes. I called this mechanism an I-am, or self-aware consciousness.

I imagined that the statement, "I am," carried within it the capacity to punch out a computer card in a special pattern which described its own identifying characteristics, its own individuality. Like the roll on a player piano, it could then play its own special tune using the harmonics of the universe. It would resonate at certain vibrational frequencies but not at others. Many of the notes found in one tune would, of course, be found in others, like the notes of a piano. But the special way in which they were arranged (or omitted) would reveal the identifying pattern of frequencies which were unique to each I-am.

Thus, I-ams are themselves creating their manifested cyclic behavior. It is very important to appreciate that cyclic behavior is not being thrust upon I-ams. It is one thing to imagine that the universe functions like a mechanical clock, or in this case, a box of player piano rolls, which have been set to run mindlessly throughout eternity. It forbids freedom. It is quite another matter to appreciate that I-ams create their very own tunes, as any composer might, from a vast array of possibilities. A study of cycles reveals that these frequency possibilities exist. But appropriately interpreted, it should not suggest predetermination.

An I-am, or the sponsoring agent of an ectoform, is able to "create a cycle" because with each blink of Laminated Spacetime, it must change slightly. It is first and foremost in the act of becoming. It is becoming slightly more open to the enabling power of the endosphere or slightly less responsive to it. If an I-am is expanding, growing, or in the springtime of its experience, it will ride up towards the crest of such activity suddenly. Its natural momentum then slows as it reaches its crest. At its peak it gathers its momentum once again but this time in the destruction of what it so recently created.

It is the multiple actions of the double helix at work (as fully explained in the physics book) but this time we see it, not within matter, but within events themselves. Events are nothing more than matter-in-motion, remember. They are ectoforms. If we fail to appreciate the integrity of a football game as being equivalent to the integrity of a fat molecule, it is because we look at things with an inappropriate bias.

The I-am is the endospheric prototype for the out-pictured DNA chain found in cellular life which is said to provide the direction for the cell's behavior, type, life-span, etc. (The DNA chain is also displayed as a double helix.) The I-am is to the DNA chain as the mind is to the brain and as the soul is to the body. The endospheric I-am is the cause of its own expression in physical form and it demonstrates itself by existing in a particular form and behaving in a particular way. This applies to atoms as well as cells. It also applies to events like a war or the preparation of dinner.

We create a war, incidentally, by having a critical number of individual I-ams come together at the creative or subliminal level to produce an ectoform called war. This is why an arms race is so likely to produce war. The attention given to arming is attention given to the I-am called war. Do this with sufficient concentration and war will have to manifest. If we should ever come to desire peace, it will first be necessary to concentrate our efforts upon an I-am called peace. We cannot arm for peace any more than we can concentrate on being sexually provocative and expect to remain virgins. It is to be hoped, therefore, that a knowledge of *how* events are created could reduce political lunacy to something less dangerous in this Nuclear Age where technology has changed our world but not improved upon our wisdom.

I-am is a statement of definition and singularity. I-am immediately defines what I am but in so doing it simultaneously demonstrates what I am not. I am a muscle cell. I am an electron. I am the Taj Mahal. I am the black race. I am a musician. I am the Vietnam War. I-am is a gestalt formed by a thought or idea, for this is the process and the only process which can define the endosphere. To borrow the words, if not the exact intent of Descartes, "I think, therefore, I am." "I am self-aware, therefore I create an I-am."

It is the purpose of the endosphere to know itself. It does this by holding up a mirror, an ectoform built around an I-am, for itself to look at. While experience will affect an I-am and thereby change it, and while mutation can and does take place as nature (the endosphere) continually experiments with the art of the possible, evolution, as it is presently imagined, is quite impossible.

The I-am of an ape and the I-am of a human are two quite different I-ams for which the ape is probably enormously appreciative. Analogously, in the creative world of cookery, a simple meal of bread and cheese does not become—through some as yet unsubstantiated process of evolution—a nine course banquet. Both are creations unto themselves and both serve their own purposes. One is not better than the other. One is different from the other— a reflection of the chef's capacity.

All I-ams—be they men or rocks—wish to express *themselves* within a range as wide as possible *for that self*. But it is sheer arrogance which makes a Darwinist think that a fish would *want* to leave the ocean, develop legs, stand erect, and grow up to be a biologist. The imagined evolutionary ladder implies lesser and greater species, inferiority and superiority. Tell that to the iguana and see if he agrees! For a fish to evolve into a reptile would be a denial of selfhood at its most basic level, for selfhood cannot be abrogated and supported at the same time. Selfhood is quantitatively much more a description of what it is *not* than what it is. Selfhood, if the concept is to have any meaning at all, must be precise in its limitations. Therefore I-am cannot become what, by its own definition, it is not.

Another way to describe an I-am is to call it an idea—an idea-of-self. It is conceptual thought and what it expresses must be in place before any creativity can commence which might effect its physical formation. The subjective idea of confronting one's employer must be in place *before* a union strike (an I-am) can find physical expression. Likewise, the subjective idea for a hydrogen atom must be well defined *before* the endosphere can produce such a quantum. An I-am is a mental blueprint.

If you pick up only one impression from this entire chapter, let it be the impression that the unlimited and undefined endosphere can only be given definition through *a pre-visualized description of limitation.* An idea about something—about what that something is and what it is not—immediately defines the undefined. All thoughts, all ideas have, then, the potential for becoming things.

*   *   *

Physical forms are built on gestalts of I-ams. The I-am of a toe bone is included in the I-am of a foot which is included in the I-am of a leg, etc. Nor does logic dictate that such gestalts of I-ams stop with us as physical bodies. Why should they? I rather imagine that our intuited knowledge that there is something beyond our own limited horizons—something rather human in its attributes, and which we usually call God—should more properly be called an over-soul. An over-soul would be a personality gestalt of which we are an individualized member.

And what lies beyond this arrangement? I suspect another hierarchy. And another. "Bigger fleas have smaller fleas upon their backs to bite 'em. And smaller fleas have smaller fleas so on to infinitum." How small we are in the flea kingdom I cannot guess. Nor does it matter. We are all one within the endosphere.

Parenthetic to our descriptive discussion of the cosmos, I would also like to make a suggestion at this point. You are indeed an actor upon a temporal stage, playing out a particular role on behalf of/in concert with a more expanded personality. The best, most productive, life will be one which is in harmony with that "super-personality." For instance, nobody wants a foot which doesn't listen to the collective needs of the entire self. Nor does the foot profit by misunderstanding its interdependence. While I would hope that you could regard this super-personality more as an equal and less as a God, you will be miles ahead of the game if you are consciously in touch with this personality. Each of us also has personal guides.

I mention these aids because I feel quite strongly that the most profound spiritual fracture which runs through our materialistic society is the feeling of alienation, of having to go it alone. It may offer you positive support to know that you have your own special system of personal concern.

Because of such an august cosmic arrangement we are all an indivisible part of All That Is while simultaneously being special and individualized. The tiny cell is itself. It has its own I-am, its own identity, even while it belongs to larger and larger gestalts. Could anything be more grand? We live within an interdependent network, attuned to the most delicate of harmonies, sustained by the most splendid patterns of cooperation and yet even the tiniest particle enjoys its selfhood. Imagine being able to put together a system so complex without doing violation to a single particle. Imagine giving that particle its own selfhood and the opportunity to explore that selfhood while simultaneously putting it into a caring context of being one with All That Is. It is inspiring to the point of transcendental reverence! Yet this is the design obscured by the preacher who represents the cosmos as a system of reward and punishment. This is the design obliterated by the scientist who declares it to be founded on a Big Bang and viable slime.

*   *   *

A second way of describing the endosphere, or the reality within, is to call it I-AM (with capital letters) or Creating Consciousness. I avoid calling it God because the common understanding of God is anthropomorphic. God is usually envisioned as a man (how quaint!) endowed with superhuman powers. The perceived relationship between ourselves and "Our Creator" becomes, then, external to ourselves with destructive emotional overtones of fear, submission, humility and guilt. Events are imagined as signs of God's approval or displeasure and the ethical codes developed by such beliefs are destructive. I-AM is All That Is. I-am (lower case) is an individualized part of All That Is as the petals of a flower are part of that flower. If the flower is divine, then so too are its petals.

The endosphere is All. It is the cause and the effect of all we know and much more besides. There is nothing but the creating endosphere so anything it creates is made only of itself. This means that the entire endosphere "belongs" to each manifested creation because of that indivisibility. If you would call I-AM God then you must say that all manifested creations are an aspect of All That Is (God) and therefore *are* God. You must realign your thinking to understand that we, as well as all else that exists, are divine. 100% divine.

It is not enough to say that there is a divine spark within each of us as if there were a small percentage of something divine and a large percentage of dross in our make-up. Most of us have been taught to equate God with good. Wishing to enjoy the comfort of a kindly, parental, and protective figure, we endow him with benign characteristics. Unwilling to compromise our image of God, we assign all the unpleasant characteristics to Satan. An unhealthy dichotomy therefore arises and life becomes a struggle between forces—forces which have nothing to do with the facts.

I-AM is creative potential. I-am is that potential given finite expression. Put it another way: I-AM is the paints, the brushes, the canvas, and the imagination. I-am is one particular painting. Only if you understand that it is the sole purpose of I-AM to express itself in all possible ways can you grasp the intent of I-AM. I-AM is amoral. It makes no distinction between flowers and weeds. Humans make these distinctions. I-AM is All That Is. This includes—unfortunately for those who wish it otherwise— the Sistine Chapel, World War II, syphillis, mother love, homosexuality, and the rose.

Now that I have insisted that the endosphere is amoral and bent on simple expression, let me pull in my horns a bit. While everything we describe as "good" or "bad" is a divine manifestation of

I-AM, I-AM (the endosphere) is heavily weighted in terms of its own success. As an instance, it is far easier to maintain health than it is to get sick. If this were not the case, we would only have a 50/50 chance of surviving any particular day. Robust health is normal. Vigorous seed production is normal. Peace is normal. Useful work and loving cooperation are normal. On a scale of 1 to 10, perhaps 8 is normal. I-AM means to succeed and life, therefore, is very positively cast in terms of that happening. If we wish to create negative I-ams, so be it, but it will take a greater effort.

The catch is that All That Is is so caring that it has endowed each of its creations with its own power to create as well as *the freedom of self-determination*. Thus, I-AM is offering everything it can in the way of perfection without violating that most precious of all legacies—the power to create self according to the wishes of that self. We have been given the freedom to use creating power as we choose to express our lives in any fashion we wish. If we have chosen adversity because of that freedom, it in no way diminishes the gift, does it? And what greater gift could there be? Imagine for a moment what it would be like to live in a "perfect" world in which you had no say in that perfection but could only fulfill your perfect destiny as a perfect cog in a perfect machine. You should be so lucky! A perfect robot!

7.

Creating Creations

AN I-AM IS AN IDEA—an idea of what the self is and how it should be expressing itself. The "self," of course, is not limited to human selves. A rock has its own idea of self. So does a pine tree. So does a traffic jam. An idea is a mental act of definition. It puts borders and edges on what heretofore was undefined. Let's return to the computer card analogy. An idea takes a blank card and begins to punch out a design, thereby describing what it means to be and even more, by omission, what it is not.

This is just what the endosphere loves. It is a creating medium. It yearns for anything which will give it definition. Then, with one more ingredient, it can put it out into physical reality. Think of physical reality as being a display screen. Some I-ams are very simple—an oxygen atom, a worm, the act of making a bed. Some are very complex—a drought, the human body, a world war. In all cases, however, such things and events are *caused* by the definition of the undefined whole through the primary mental act of forming an idea. This is not in question. Who or what is doing the thinking is the only question.

In order for an I-am to manifest physically, an I-am needs only one thing more than the conforming guidance of an idea. It must attract to itself what I have come to think of as a critical mass of intention. The single difference between a simple idea and an

I-am which becomes a physical manifestation is that an I-am is an idea accompanied by a sufficient amount of intention to give it physical form. Every physical thing or event is, then, an idea which has been nourished into physical being by intention.

Intention can also be described as determination or will or conviction. Intention is very emotional in its character. Intention carries the meaning, "This is *true!*" Seriously held beliefs, for instance, are automatically I-ams. They are ideas which have been wrapped in the conviction of intention. They create themselves as events.

To illustrate this "critical mass of intention" and how it works, let's look at two neighbors. Each thinks it would be a good idea to build a house on his property. One dreams of that house—has an idea about a house—but lacks sufficient intention and the house never gets built. The other has the same dreams but he adds intention to his idea. The more intention he adds, the sooner the house will be built.

Think hard on this example. Do away with the notion that it is physical energy directly which gets the house built. Physical energy is the by-product (the effect) of intention, for at any moment, if intention fails, the house will not get built. Intention is the mental cause of physical energy. And physical energy is *the effect* of such a cause. In order to get the house built, we need x units of intention. Intention will automatically supply the project with the requisite man-hours necessary to bring the I-am of a house into physical being. It may take a million units of intention to complete a house and only one unit to cause a leaf to fall but the process is the same for both events.

For practical purposes we should think of intention as an energy complex. At the same time, we should also clearly understand that intention is a mental capacity. It is the enabling power of the endosphere we speak about when we speak of energy. This enabling power moves things and events towards a likeness of the guiding idea behind a particular I-am. All "energy" is generated from a mental level. It is behind the phenomenon of cyclic behavior. By the time we see this energy in physical form it becomes the dimension of time, which includes chemical, electrical, atomic, etc., as well as muscular. (The equivalence of energy and time is discussed in the physics book.) The properties of the various materials used in "energy-producing" ventures are *subjective* properties, born of the intention of the various molecules and cells to behave in a particular way.

Consciously stimulated or unconsciously inspired, intention is behind the quest for knowledge, the orbiting of a planet, the bloom of a flower, a lover's kiss, a murder, the action of the ocean's tides, disease, the creation of a hydrogen atom, war, and peace. Physical reality is the out-pictured expression of I-ams, which have been fueled by intention sufficient to make them "real."

*   *   *

Getting a house built by the use of intention is not difficult to grasp. Not only are we used to thinking of house building as a creative act, but we can see exactly what is happening as the house is being built. But an auto accident is also a creative act and it too is created in exactly the same way. The definition of an idea is given life, or sponsored into physical reality, by a critical mass of intention.

One difficulty in recognizing an auto accident as a creative event is that science, which would advise us on all matters, has avoided, with considerable skill, giving the mind any serious consideration. Scientists prefer to address themselves to the brain, which is nothing more than a computer and has nothing to do with the skills of the mind. The reasons for this failure on the part of science are multiple. For one, the mind, as opposed to the brain, does not submit itself to a laboratory in a way which makes scientists comfortable. The methodology of "objectivity" has long since become the *sine qua non* of the scientific community. It is more important than truth. So they have cut their truths to fit their methods as if a tailor were prepared to make only one style of suit for all time. In terms of methodology, scientists wear 17th century garb.

As the cosmos is created from a subjective level, all of its manifestations are subjective or emotional/mental in origin. The scientific attempt at objectivity is therefore impossible, for even the decisions on what to test and how to test it are subjective and the results will only describe the test, never the tested. The cutting edge of the scientific community is beginning to recognize this with such questions as "What part does the observer play in what he is observing?" Even so, there is no stampede to challenge the very foundation of science—it's methodology. So, for the time being, scientists are stuck with examining and evaluating various effects. They never touch cause. They believe in their objectivity, but it is only an illusion.

Another reason that the true "cause" of an auto accident remains so obscure is that it seems to be human nature to avoid blame for events we do not like. Sickness, mayhem, or the loss of a job are events we like to consider accidental. They seem to happen "out of the blue" because we prefer it that way. Conversely, we are much more willing to accept responsibility for our good luck. If we marry the right woman, get an award for excellence, make a quick bundle on the stock market, we are much more willing to attribute these "accidents" to our own special powers. In terms of *how* good or bad luck is created, however, there is not one particle of difference.

*   *   *

The best way to look at what goes on in a mind is to think of the mind as a potting shed. In this shed, some of our ideas are but tiny seedlings. Some are all but ready to be transplanted into physical being. Every second of our lives we work with the seedlings in this potting shed. We either encourage the seedling ideas by giving them our conscious or unconscious attention or we let them die. We nourish them with our emotional responses to life around us. "He's not going to get away with that!" is an angry response filled with lots of intention. So is, "I'm sitting on top of the world!" That belief expresses the feelings, and therefore the intent (and therefore the experience), of personal satisfaction.

Fear is, by all odds, the most intent-filled emotion we experience. Look at what privations we are willing to go through in the name of our fear of Russian conquest. Look at the non-renewable resources we squander, our willingness to let some of us go hungry, and our choice to accept illiteracy in 20% of our population, so that we may arm ourselves.

We imagine a dreaded outcome with great intensity. While this allows us to move to safer paths if the fear is justified, fear, nourished in the potting shed of our minds, continues to express itself in I-ams, thus actually producing in our experience what we fear. When we are afraid our ideas are sharply focused. What we fear becomes real mentally and when it has been made real on this level it has achieved "critical mass" status with the power to convert to a physical event. There is no such thing as an unexpressed idea if it is accompanied by a critical sufficiency of mental energy or intent. It *must* manifest.

It should be understood that each thought seedling matures to represent itself. Petunia seeds don't produce roses. In the same way, a belief (an idea wrapped in emotional commitment or intention) in violence—whether from the fear of it or the espousal of it—can only produce violence, for to fear violence is to make it real in the same way that faith in its efficacy makes it real.

In human terms, physical reality is nothing but the out-picturing of our beliefs or that which we create because we have mentally nourished an idea into an I-am. Physical reality is the symbolic display of inner reality. We do not think of the act of nailing one board to another as symbolic but it is. It symbolizes the intent to build a house. So it is with an auto accident, though exactly which of our many thoughts it is symbolizing may escape us because we have grown used to finding cause in other places. We find cause in the road conditions, the other driver's intoxication, a mistake in judgment. Yet these are only the stuff with which we express our beliefs in the same way that our house builder uses nails and concrete and wood according to the design he has in mind. The difference between the two is that house building is an overt and conscious act. Creating an auto accident is "unconscious" in that we are not paying attention to cause. We think it's okay to have destructive thoughts as long as we do not "act on them." But if we do not act upon them consciously in ways which channel this energy, they will act for us. Every auto accident is the creative act of two or more people who believe in being victims.

It might be helpful to point out again that in the greater reality in which we dwell we are one with All That Is. It is therefore as easy to arrange a rendezvous with a drunken driver who is also looking to create an accident as it is to get one's fingers to pick up a pencil. Imagine, if it helps, that the endosphere is a big computer which matches up computer cards in such a way that the creations of all participants will dovetail in the same way that the entire body becomes attuned to the act of picking up a pencil.

If you are interested in a more detailed explanation of how events are created, you should read another book of mine, *As You Believe*. Not only does it go into this subject in depth, but it explains exactly how we can eliminate negatives from our lives and move on to the conscious creation of "miracles."

Because we are, for the time being, physical, our capacity to create has its limits. We are able to handle only so much endospheric power at any one moment. This is why a house takes months, perhaps even years, to build. Without this restriction upon our capacities, we could summon sufficient intent behind our ideas to make them manifest immediately. In that case, the house would just suddenly appear without any "work" being done on it. It could be accomplished as a finished product in the mind before any of it manifested.

No human in a body can approach this kind of expertise. Even Jesus, one of our great adepts, had his limitations because of his physical restrictions. On the other hand, we are far more capable than we presently believe. The art of *consciously* summoning intention in order to perform the creative skills of which the mind is capable lies, as yet untapped, in most of us, principally because we have not been introduced to these possibilities.

I have been "healing" psychically for many years. I have had many notable successes. I do it all by generating sufficient intention. After that I just let "nature take its course." My entire thesis is that we do this naturally every minute of our lives, but are as yet ignorant that we do so.

I consider there to be only one difference between myself and anyone who claims he can't perform such miracles. (A miracle is a positive event which appears to be unusual because of a lack of information.) I have taken the trouble to investigate the possibility and to experiment until I achieved success. I have paid my dues like any other performer. It takes effort to become an olympic competitor or a concert pianist or a surgeon. It also takes effort to become a healer. Those who display these gifts "spontaneously" have probably honed those skills in other lives which only adds to the mystique which surrounds this sort of talent. This is unfortunate for we could all be conscious healers if we wished.

* * *

Because we have concentrated so determinedly upon physical experience as the end-all-be-all of human experience, we have given little attention to the true cause of life. We mistakenly believe, therefore, that we are at the mercy of life rather than its creators. Such beliefs make us feel impotent and we have hastened to fill in for these perceived weaknesses with technological aids. We are not encouraged to use our natural telepathic capacities. We have phones. We do not need total recall. We have computers. We do not need our homing instincts. We have maps. We do not need to practice health. We have doctors.

I suspect, for instance, that the pyramids and Stonehenge were "houses" which were built, largely by the use of mind power—a power which appears to have long since atrophied in us for lack of use. My guess is that the stones were mentally levitated into position. I also guess that these same stones were hewn to their reputed accuracy by using that same mind power in a different mode.

Book after book has been written remarking upon some of the capacities of the human mind. If their wisdom hasn't permeated the intellectual mainstream of our culture it is because we have chosen to take our lead from scientists who belittle such phenomena because these phenomena lie beyond their own self-described perimeters. Uri Geller bends spoons without touching them. Others control the flip of a coin with 100% accuracy. Yogis can manage their vital signs through mental discipline to the point where they are pronounced physically dead. Aborigines can walk on red-hot coals without suffering so much as a blister. There was a blind woman in Russia who could "see" color through her fingertips. Another man can project his mental images of what he has just seen directly on to film as if he were the camera itself. The list of "miracles" is large and begging for appreciation and investigation.

As thinkers, I do not know how far down the totem pole we belong. For all I know the entire solar system is but an atom in the body of a sapient giant whose size surpasses our comprehension. I *do* know that there are minds who can create galaxies in a single blink. They do not need to wait upon their intention to unfold itself because they are not limited by a dimensional reality.

Not only can these minds create galaxies which appear to have a particular "age" to them, they can, should they choose, materialize or de-materialize these same galaxies. I say this somewhat glibly. But if I can do this on the small cellular level of the body, the "miracle" of such possibilities can already be demonstrated. All that is missing is an awareness that infinite possibility is far greater than finite possibility. At one level we belittle ourselves. On another, we aggrandize ourselves by believing that we are at the top of the heap.

For these reasons the scientific statement that the earth is 4.55 billion years old (a figure which constantly gets larger) with the implied assumption that it didn't exist before and will cease to exist at some moment in the future, are the products of simplistic thought. The earth, or the idea of a place to experience physical reality, is eternal. And do not forget—we get a completely "new" earth with every blink of Laminated Spacetime. A brand new earth in "old" form.

While we are discussing beginnings and endings, perhaps a word on death is in order. The Theory of Laminated Spacetime confirms the eternal existence of the soul and temporary existence of the body. The Creating Cosmos demands that all births and deaths are self-created for reasons known best to the creating entity. Accidents do not happen in a physical reality sponsored by the Creating Cosmos. This is going to make for a lot of resentment on the part of those who cherish the innocence of a murder victim, a cancer patient, a young war hero, etc. We like the image of the innocent and the guilty. But as the cosmos is founded on the freedom to create what we wish to create, birth and death must also be of our choosing. We cannot be victims of life though we may choose to express ourselves as victims of that life as a belief in that truth. A murder represents the coming together of two or more people, each with dovetailing, self-created needs. One is fulfilling his need to be a victim, the other is fulfilling his wish to express contempt. They are a match-up on the computer. When society no longer produces self-described victims, there will be no more mayhem and very little disease. To rid ourselves of crime, it is not the criminal who needs our attention so much as the mindset of those who need criminals to fulfill their needs.

In the true reality of All That Is, there is no distinction between "them" and "us." All is One. To define some minds as superior and to think of those minds as separate from us is not particularly appropriate. On the other hand I wish to underscore the fact that mind power has a far greater capacity to create than anyone in a *physical* body can achieve.

The *idea* for a human comes from a mind unrestricted by physical circumstances. We are the "thought-children" of such minds. Or you could say it this way: we, as such minds, thought ourselves into finite existence. The best analogy I can think of is that a super-thinker can be likened to an actor who chooses, for the moment, to take a part in a particular play. He is not less because he struts his time upon a stage but neither is he limited for all time to playing Hamlet. While he endows the role of Hamlet with his entire, through momentary, being, he is greater than one particular role. In that sense, he is a super-thinker.

And this is why I say we are creating creations.

8.

Multiple Realities

I WOULD NOT HAVE THOUGHT of multiple realities myself—at least not to the point of including a discussion of them in this book—but I have read the Seth works transcribed by Jane Roberts. The entire concept seems not only eminently logical, but indeed mandatory. The Seth books go into detail about multiple realities, probable selves, and much more. If you are at all interested, you should read them.

Nothing in nature suggests singularity. We have birds, trees, atoms and cells of every description. Is it reasonable to assume we would have only one reality? Furthermore, is it reasonable to assume that all realities should be built on the principles arising from spacetime? If you were the Great-Potter-in-the-Sky, would you want to make only one pot and be done with it? Potters I know just don't do that. They experiment with every possibility they can imagine.

For me, the unlimited potential within the endosphere *demands* that it be "siphoned off" by the creation of uncounted realities. It therefore occurs to me that when one cycle is up in this reality, it may be down somewhere else. For instance, at the time of this writing we are going through what is euphemistically called a recession by those who have kept their government jobs. I keep wondering if some other reality is enjoying full employment and credit card mania the way we did a few years back. If they are as ignorant as we of the economic seasons, however, they will be no better off than we are presently when we start our ascent out of the pit and they commence their descent into it. And what of nuclear explosions? When we take more power than we are normally "entitled to," does that cause a dearth somewhere else? Did we perhaps run out of power economically because in some other reality they were draining off our energy? Of course, I don't know the answers to these questions, but it makes for interesting speculation.

Laminated Spacetime can, however, provide the mechanism for multiple realities. Look at it this way: telephone technology is set up in a way that a multiple number of messages are blipped simultaneously over a single wire. (And if mortal man can do this, why not He Who Makes Trees?) Let's say that only four messages are being transmitted. The message units would look like this: A B C D A B C D. Each conversation is broken into bits and synchronized with the other conversations in such a way that nothing appears to be missing. Nor can we hear—usually—any other conversation. We hear only A A A A. But very occasionally we do hear B B B B in the background. But then, only rarely are ghosts reported or fairies sighted. Only rarely do entire armies disappear without a trace or a squadron of airplanes de-materialize as one did not too many years ago in the Bermuda Triangle. Only rarely are UFO's reported.

I am not suggesting that multiple realities are the answer to any of these phenomenon, but we should keep an open mind. Personally, I think ghosts and fairies come from another layer of *this* reality but we are normally blind to them because of our visual conditioning. As for those wholesale disappearances, who knows? My mind takes easily to imagining both "flaws" in the blinking process and the warping of the spacetime webbing by nuclear destruction which might cause a skip in the beat and thus what was materialized out of this reality at the end of a blink might very occasionally fail to rematerialize here but materialize somewhere else. I also find it easier to think of a UFO as coming from another earth reality which has already solved the puzzle of trans-reality travel than to think of the UFO as coming from another galaxy. One thing is certain: we cannot even begin to explore such possibilities if we insist upon describing our reality as the only reality. It is not my purpose to provide any sort of answers to these puzzlements. My only purpose is to get your imagination working. But in making any judgment of my suggestion about multiple realities, it is best not to limit Infinite Capacity to mortal credulity.

To myself, I describe spacetime as organic in its structural characteristics. I imagine that realities look something like a spoonful of caviar or an amorphous cellular mass. Capable of "cellular" division, realities would build up or die back like all cellular constructions. Seth calls the selves in some of these realities probable selves for the self is not limited to a single reality either. Only our specialized awareness, limited in great part by our training and lack of imagination, makes such possibilities seem unlikely.

Think about multiple realities. It won't hurt you and it could be fun!

9.

The Purpose of it All

IN THE FINAL ANALYSIS, I don't suppose the Creating Cosmos has a purpose greater than the joyful expression of creative possibility. Solely in the service of that purpose it is a design of the most sublime construction. It is breathtaking both in its simplicity and its opportunity. It grants total freedom within a context of total cooperation and partnership. There are no winners and losers in the Creating Cosmos concept. Because each plays a game of his own choosing there are only winners.

Human consciousness, however, wants more for itself than the experience of simply being. It wants to master its craft. Not so much, I suspect, because it prides itself on craftsmanship, but because it is desperate to put as much distance as it can between itself and life's sorrows. To that end, then, there are several clues implicit in the design of the Creating Cosmos which can bring us closer to those goals of happiness and "right action." While I recommend *As You Believe* if you are more than casually interested in improving the quality of your life, there are still some obvious remarks that are germane to our current discussions.

It seems apparent to me that physical reality has been invented as a demonstration reality. Let's call it a workshop. In the greater reality of the endosphere in which we more truly dwell, it would be too difficult to comprehend what we are doing with our creativeness, because all thoughts would be immediate and composite. Without the limitations of space and time, no-thing would be happening all at once. Did that confuse you? Without space to put a thing into, a thing would have no dimension. It would therefore be nothing. No-thing. Without an object's ability to endure from minute to minute in a process of change, it would have no beginning and no ending. The experience of time would be a composite of all the events which happen "in time" all jammed up together as if the pictures from a series of negatives has all been printed on top of one another.

The creation of ectoforms through the laminating of spacetime permits us to live concretely and slowly within our own acts of creation. The advantage of physical reality, and hence its implied "purpose," is to give us an opportunity to experience what it is we wish to create. In the endosphere cause and effect are identical because there is no time (duration) and space (extension). In physical reality, timelessness is stretched out so that we may more easily discover the internal connections between ourselves as creators and the experience we have chosen to make real. We are permitted the privilege (or distress?) of living within our own thought patterns.

As the potter cannot understand the art of making pottery until he has worked with clay and appreciated the results of his work, so we cannot understand the consequences of our own creative capacity until we have lived within our own created acts. This, then, is the implied purpose of life—to learn about and experiment with the use of our great creative power. We should expect from ourselves, I think, that the day should finally come when we awaken to this game plan. We should eventually discover from whence the ill wind blows. Ourselves. Given that insight, it would be reassuring to hope that we might wish to re-address ourselves to the task of enhancing the quality of life throughout physical reality. That particular choice, however, is ours, for choice is the blessed and undenied gift of a design which gives creative thought its due.

In terms of a commandment, I suppose I would put it this way: *Thou shalt not whine.* If ever we were to quit blaming others for our misfortunes, two things would immediately happen. First, by taking responsibility for our acts of creation we could come to understand how profoundly powerful we were. Those who take responsibility are aware of that power. They recognize consciously that they are sourcing their lives. There is nothing quite as satisfying as being in touch with this knowledge. Those who blame others have chosen to play at being the victims of life. They feel impotent. But remember, no matter how miserable life may be, if you did it you can undo it, and just knowing this is 85% of the battle.

Secondly, one half of the population could quit trying to make the other half of the population feel guilty. People can be destroyed by guilt and it's the guilt throwers who are the hatchet men. Yes, those who accept that guilt are spinning their own webs, but guilt throwers themselves are only externalizing their own feelings of guilt. Guilt is like an enormous beach ball which, once it is brought into being, gets tossed around until somebody accepts it and feels awful. That's no way to live. With more of us taking responsibility for our own acts and feelings, we could begin to move away from the game of me-OR-you which molds most of our personal, political, social, and economic relationships into the far more profitable game of me-AND-you in which everyone wins. We might even reduce our lust for revenge, our preoccupation with competition, our intolerance and bigotry. Instead, we could concentrate on the work of art we are personally responsible for—ourselves.

*   *   *

As the introduction of time permits the attenuation of events so that there is the illusion of a separation between a beginning and an ending (between cause and effect), the introduction of space allows for the attentuation of the endospheric One into the illusion of the many. Imagine the cells of a single body blown to the winds and settling down to become the earth and all it supports. Some of that body demonstrates as ocean, some as plant life, some as human beings, some as war material, some as music, yet it remains a unified body. It is interdependent throughout all its parts and functions in the way any body would be, even though we speak of kidneys, arms, and eyes as separate identities.

In this expanded form, there is the illusion of separateness but it is only an illusion as ecological studies and our own intricate economic interests are already beginning to make pragmatically clear. Harm the river and we have poisoned the water on which our lives depend. When the wages in Hong Kong are a dollar a day, it directly affects the well-being of all garment makers throughout the world. We live in times which are almost mystical in their clarity. They present a spiritual truth which is not easy to perceive in simplistic cultures where resources are "infinite" and cooperation a choice.

The One that is the endosphere makes its appearance as ectoforms in the same way that white light, when diffracted, creates many different colors. Or return to that geometric dot, understanding that with every blink we dissolve back into that dot. Before and after that timeless moment we live in the illusion of separation. If physical reality seems more real to us than the indivisible One, it is because functioning in a physical world is difficult. We feel the need to focus attention upon it. Additionally, oneness is very threatening to those who are not very sure of their own identities in the first place. Oneness blurs the lines of self-identification.

One of the most important purposes of dreaming is to allow us a release from the necessary concentration on physical reality. While rest re-tunes the body for the next day's activities, so sleep and its dream life permits the mind to become more sensitive to

Oneness. This Oneness is why dreams can appear so chaotic. Time and space and other agreed-upon symbolism have less hold on us in the dream experience.

Because we are all one in the creating, sponsoring "true" reality, brotherly love and concern for all becomes the highest priority of pragmatic self-interest. Tending the aches and pains in any one part of the body improves the general well-being for the entire body, doesn't it? For selfish reasons we owe it to ourselves to make it good for every aspect of the attenuated body of One. Popular thought suggests that the fit are entitled to survive against all comers by dint of their cunning. Yet the "cunning" which permitted our auto workers the highest wages in the world is now throwing them out of work. To make such artificial distinction between two aspects of One is like saying that we can improve conditions for the heart if we force the lungs to stop breathing. To diminish one life is to diminish all life. The One Life. With the Theory of Laminated Spacetime, this age-old insight becomes a "scientific fact."

> "No man is an island, entire of itself; every man is a piece of the continent, a part of the main, if a clod be washed away by the sea, Europe is the less, as well as a promontory were, as well as a manor of thy friend or thine own; any man's death diminishes me because I am involved with mankind; and therefore never send to see for whom the bell tolls; it tolls for thee."
>
> —John Donne, 1610

For me there is no greater empiric demonstration that we are bound together in the warp and woof of a single fabric than the harmonic behavior revealed by the study of cycles. My father discovered hundreds upon hundreds of these cycles, yet how many more await discovery? His young discipline suggests a high thread count. As an illustration of what I mean, the 5.9 year cycle mentioned earlier not only appears in business failures, copper mine production, grouse abundance and sunspots, but in 17 other reported series of statistics—Lake Huron water elevations, oat production, steel production, and bond yield, for instance. Think what might be revealed if we kept data on all fields of activity. Serious marital disputes? Name preferences? Gambler's luck? The conclusions are inescapable. We are in close subliminal touch with one another.

There is such commonality in these subliminal pulses that it infers that thoughts are "catching." A single idea in the abstract becomes the guide for private but multiple expression. For instance, women, living in dorms or other forms of communal life, tend to synchronize their menstrual periods. Again the rhythmic and predictable ups and downs of the stock market demonstrate that we tend, like sheep, either to rush out anxiously and sell our stock or, with equal determination and obedience, rush enthusiastically to buy them. One year we emphatically reject Barry Goldwater for president only to embrace him in the form of Ronald Reagan some years later. At predictable times (predictable if you know your cycles) a minor incident can set off a war of global proportions yet in the trough of that cycle, Florida could be towed out to sea with little more than mild surprise on the part of the general public. We march, I believe in cadence to an inner drumming which vibrates in cosmic harmony. And we march together.

10.

The Law of One and the Cosmic Boomerang

THE LAW OF THE COSMOS states that all aspects of itself must be equally honored because it is an indivisible One. Cosmic law cannot be disobeyed without consequence, but before we entered the nuclear arms race civilization could survive its disobedience to this law. This is no longer true. If we do not learn that we are indivisibly one with all else and demonstrate this wisdom politically, economically, socially and environmentally we will—there is no other word for it—screw ourselves out of a habitable planet. We are playing in a big game with big odds and, unhappily, we don't even know the rules of the game. I therefore want to explain the rules and demonstrate how they work using contemporary society to illustrate what I call the cosmic boomerang.

When, in our muddleheadedness, we set ourselves apart from cosmic law, we will be opposed by the cosmos, not as retribution for our sins but because we run counter to the system. If you stick your finger into a flame you are going to get burned, not because the flame has it in for you, but because you have ignored the laws under which a finger and a flame function. It is you who create your own punishment in both systems of law. There is no cosmic judgment in either case. It is not a matter of good over bad, right over wrong. Cosmic law is like your elbow. You can bend your elbow safely in only one direction. If you try anything else it will hurt—badly. Likewise, cosmic law cannot be violated without painful consequences.

I want to say this again because there is very nearly a unanimous belief to the contrary. *There is no dichotomy within the design of the cosmos itself. There are no forces of good and evil.* There is, however, the freedom—for humans—to work within the Law of One or to challenge it. If we work within the law, observing the sanctity of All That Is, we have aligned ourselves with the reality of the cosmos and we will be in a state of grace. On the other hand, if we do not observe the sanctity of All That Is we create misfortune *for ourselves*. This is why I call it the cosmic boomerang.

We violate the Laws of the One when we do not give the same value to all life that we give to our own. This respect for life is built into the instinctual patterns of animals. But because humans have been given the creative opportunity to build their own "instincts" through the formulation of belief systems from their truths (see *As You Believe*), this "instinct" is not automatically part of the human system. It must be taught. Yet, respect for All That Is is *the restraining mechanism/belief* which opposes and holds in check berserk and lawless behavior. We call the degeneration of respect for life criminal behavior, and those who evince this kind of behavior criminals. When this restraining mechanism breaks down, or has not been adequately instilled, we have the kind of depraved behavior for which the human species alone is famous. To date we have not been sufficiently aware of the importance of this law to give it more than a philosophical nod. Besides, we have not wanted to limit ourselves to an obedience to it. But this was before the advent of nuclear weapons and toxic chemicals. Now we have run out of options.

We, as a society, have carefully arranged our laws—and hence our behavior—not to conform to the cosmic law of respect for All That Is, but to conform to our own self-interest. Criminal behavior which disadvantages us we condemn—murder, rape, burglary, embezzlement, car theft, terrorism, etc. Criminal action which advantages us we approve—war, the use of animals for medical research, the decimation of the whale population for profit, the abrogation of civil rights, criminal punishment, terrorism when accomplished by *our* government, etc. In each of the above instances the premise upon which permission for these activities rests is that some life is better than other life. Yet such a premise is in defiance of cosmic law—a law which states that all life is to be revered, not just American life, not just the lives of good people, not just human life, *but all life*.

*   *   *

I would like to show you how the cosmic boomerang works. The boomerang is the cosmic reaction to your inappropriate action. What you put out comes back to you in the same but symbolic form. I'm doing nothing so impractical as making a plea for moral behavior. Such pleadings are futile. This is a lesson in pragmatic self-interest. If you don't like being hit in the face with a horse puckie, here's your chance to learn something other than how to duck.

The cosmos is set up so that if you depreciate an aspect of the One Life, you will make an enemy of that which you depreciate. For instance, suppose you call the Russians an evil empire and jokingly announce that you intend to nuke them in five minutes. Because of your disobedience to cosmic law, you have just produced an enemy that is bigger than he was before you made those remarks. On the other hand, suppose you send the Russians grain because of your recognition of their humanity. You have, by that act, made a bigger friend. In the first instance you are out of sync with cosmic law and it will boomerang by making an enemy for you. In the second instance you are in harmony with cosmic law and you will profit by your acts. Whenever we use a cosmically criminal mentality to find a solution to a problem, we will be stuck with the consequences of a self-generated "punishment."

Because we have, over the centuries, disobeyed the most fundamental of all cosmic laws, and because those laws cannot be disobeyed, these laws are putting the squeeze on us. In earlier times we did not have the expertise to create toxic chemicals and radioactive wastes. Economic structures were folksy backyard arrangements, not globally sensitive industries. War was a neighborhood affair and limited by technology. If we have been blind to the one world concept in more simplistic centuries, it is confronting us head on now.

Let's look at the boomerang effect as it applies to international policies. I want to show you, if I can, why facing a world-wide nuclear Armageddon is the lawfully appropriate consequence of our transgressions against cosmic law.

We are a Christian nation, imbued with the Christian ethic. (Russia shares these roots in spite of the present public policy of atheism.) Although the Bible frequently exhorts its readers to true cosmic atunement, its interpreters have concentrated their efforts on the saints-and-sinners message, it being easier to understand and more satisfying to expound upon than loving one's neighbor. The Crusades, the Inquisition, the Holocaust, and other missionary efforts have therefore typified Christian politics. The ethic, by creed as well as by example, teaches that some life is good life and some life is bad life.

While almost all nations are active participants in a disregard for the cosmic Law of One, the United States, as the most favored nation, has been the initiator as well as the model for world ethos since the end of World War II. And we have been sticking it to the Russians ever since we knew we could win that war and no longer needed them as an ally. It is a matter of record that we hastily dropped the atomic bomb on a Japan which was already suing for peace, not because we wanted to hasten the end of the war but because we were afraid it would end too soon to put Russia on notice that she was, without any argument, number two dog. The atomic bomb was used, not to complete that war, but to start the next.

We have had it in our power these past decades, by virtue of our many advantages, to make *our* policy international policy. We *still have that power.* I am therefore not as concerned about the iniquities of other nations. I am concerned about us, for where we lead others will follow.

Until the end of World War II we considered ourselves to be a neutral nation (except in South America), preferring to mediate between nations than to involve ourselves in ancient squabbles. All that changed when we became, indisputably, the most awesomely powerful nation in the world. Such power corrupted us. Instead of casting ourselves as the world's good guy, our first concern became our own power base. And so instead of hemispheric domination we went for world domination. As a result of this change in policy and 40 years of effortful skullduggery, we are, internationally, never really trusted and usually hated. It is the direct result of using others to fulfill our megalomaniacal purposes. (To use others for your purposes is to state to those others that your purposes have a greater value than theirs, remember. It guarantees a counter-reaction.) In short, we imposed our ethic on a world already disposed to accept it. It is appropriate, therefore, that any denouement also be world-wide.

A nuclear missle has two important and symbolic aspects to it. These aspects make *nuclear* annihilation totally appropriate to our present course because the cosmos can speak only in symbols. The first aspect is obvious. A missile directly threatens life. It makes a sham of peaceful intentions, no matter what the rationale. Its *raison d'etre* is to annihilate life. We cannot build bombs and respect the cosmic Law of One at the same time.

The second aspect is less obvious, but is perhaps even more important symbolically. At the base of nuclear technology is the destruction of matter. You may not consider matter to be "life" but matter is an indivisible component of the cosmos when it manifests physically. More than this, it is the building block which makes physical form possible. If you read *The Theory of Laminated Spacetime* you will find my arguments challenging the belief that nuclear destruction is taking place on the sun. Nuclear fission is *not natural*. It is a man-made technology in terrible contravention of cosmic law. Therefore having defied cosmic law at its most fundamental level, the boomerang has been activated to respond at the same most basic level—total annihilation. What we sow, we will be forced to reap unless we can manage to get things turned around.

The same mentality which breeds a collective nuclear arms race breeds the individual street criminal. Cosmically, our disregard for the truth that all life is to be respected equally swings around to "punish" us through those who are least able to resist popular attitudes about the worth of another life. The street criminal may be society's weakest link but society has, nonetheless, given him permission to think his thoughts because of its own attitude about life. (I hope you will do yourself an enormous service soon after you have read this chapter. Turn on TV and watch any popular cops and robbers show—the more popular and lustier the better. You will be newly shocked by the principles on which our society is founded. Good guys *should* waste bad guys. As all of us think we are good guys, the lesson to be learned from our popular moral scripture (TV) is, "Disregard everything but your own cause!"

In a democratic society, the government is the out-pictured reflection of the wishes of the people. By looking at our government it is perfectly obvious that we approve conquest, subversion, sabotage, murder, entrapment, etc. We ask our government to do for us collectively what we wish we could do individually. In other words, if it is the wish of the consensus to behave immorally, the government will reflect that wish in its own activities. It takes no great insight to see that if our government were an individual and not a consensus it would be put away in an institution for the criminally insane. Therefore if you want to know why crime mounts and becomes more bizarre in its detail, look to the government we have elected and encouraged to become our example in its immorality.

Governments, unfortunately from their point of view, have an obligation to be more than a reflection of society because, in circular fashion, they are something more than a collective fantasy. They are models, presumably managed by the best of us. So, until governments set a good example, we will follow their bad example. Our rising crime rate is an exact reflection of a government which increasingly fosters criminal acts. We cannot argue with the cosmos. Our collective lack of respect for life will automatically plague us with the consequences of such a belief in the form of muggers and sodomists. Because we believe in the inequality of life, this cosmic error will create its own backlash in increased street crime.

*　　*　　*

There is a direct correlation between our approval of the use of animals for medical/scientific purposes and the alarming rise in teenage suicide, between the accelerating incidence of AIDS (acquired immune deficiency syndrome) and child abuse, between the pollution of our environment and wife beating, between a futile penal system and the devastation of the whale population, between labor/management abuses and illness. The root assumption behind all these phenomena is the cosmically disobedient assumption that there is a hierarchy of greater and lesser life and that the higher one is stationed in that hierarchy, the greater the privilege. The determination of this hierarchy is based almost exclusively on the power the greater is able to exert over the lesser. Such a power hierarchy is given the hallowed position of law by Darwinism and it is in direct contradiction to cosmic truth.

A philosophical adherence to Darwinism therefore leads us very directly to the consequences of the boomerang. No wonder we've been doing so poorly!

There *is* a hierarchy within the cosmos but it is a hierarchy of *responsibility* far more than it is of opportunity. As a culture, we grab at the opportunities and refuse the responsibilities. Cosmically, the hierarchy is built on more expanded rings of consciousness which include within their boundaries lesser rings for which the expanded rings are responsible. These rings of consciousness are most easily appreciated in the analogy I have already used— the cell within the toe within the foot within the leg within the body, etc. Here we can experience first hand the cosmic interdependence of all rings of consciousness as well as what a disregard for "the lesser" does to "the greater." If your toe is abused or damaged, the entire body suffers in direct proportion to that injury. Please note again that a greater ring of consciousness *incorporates into itself* the lesser ring of consciousness. It is a matter of *expanded* consciousness, not rank.

In the body of our greater environment, our sense of hierarchy might be perceived as commencing with inanimate material and running through plant and animal life to human life. Within human life, the hierarchy continues with children at the bottom, followed by women and topped by men. There are, of course, all kinds of other imagined hierarchies based on social ladders, ethnic cultural positions, economic advantages, etc. These concepts evolve directly from Darwin's theory.

It is not my intention to discuss all of these hierarchies but rather to give you a *feel* for the consequences of such a cosmically criminal point of view. Principally, I want to explain the boomerang effect so that you can more easily understand where *self-interest* lies.

In search of the good life we have approved the manufacture and use of toxic chemicals without a heed to developing methods of neutralization. As a consequence, it is now estimated that 1/5 of all our drinking water is contaminated, thereby directly reducing the quality of life we set out to achieve in opposition to cosmic law. The cosmic law requires that if you poison the environment, you will be poisoned by it. If you violate one ring of consciousness, its "pain" will be felt by you.

Our position in relation to animal life should be that of the shepherd. For a number of reasons the boomerang effect of animal abuse is more difficult to document. Poisons in the water can be measured. Poisons in the heart can only be intuited. Our intellectual understanding of cause and effect would be better served if armies of little laboratory mice marched forward to bite the hands of their persecutors, but as they naturally live their lives in cosmic harmony they do not retaliate. They understand intuitively that we hurt ourselves more by harming them than any harm they might invent on their own. Perhaps they don't want that karma! Animals do elect to absent themselves from our universe by dying out as a species, however, and we are the poorer for the lack of diversity they could have provided.

Those who abuse animals by caging them, causing them harm, or otherwise misusing them are, in fact, directly harmed by their own acts. They sacrifice a bit of their humanity with each act because they must enure themselves against their instinctive knowledge that they transgress. They become more the robot and less the human as their acts become easier. Killing an animal for food is not an abuse if it is done in appreciation of the gift as, for instance, the Eskimos do, for the fox must have his hare. It is when we assume it is our right to kill because our interests come first that we invite unpleasant cosmic responses, not from our deeds so much as from an inability to love and appreciate.

I very seriously wonder if we are not "punished" for our approval of the use of animals in medical research by an *increase* in medical problems in the same way that toxic chemicals (in the name of better living) ironically reduce our standard of living. As health is generated from a state of cosmic harmony, so disease is a symptom of misalignment. If we appreciated all life as our own, what direct effect would this love of all life have upon a general resistance to disease? Is the mind set which permits animal abuse the same mind set which produces disease? I think so.

It is also interesting to speculate how an obedience to cosmic law would have affected our general level of expertise. What if the use of laboratory animals had been outlawed? Would scientists have been forced thereby to study health instead of the biological processes of disease in animals? By our concentration upon the dark side have we given that dark side power? While we cannot calculate alternatives never taken, it is patently obvious that our health problems have increased geometrically under the system we presently endorse in spite of the diligence of scientists to the contrary.

In another irony, the same belief which allows us to abuse animals simultaneously forces us to abuse our children. So while the boomerang effect of animal abuse cannot be statistically proven, the lack of concern for lesser rings of consciousness dependent upon our protection and the consequences of such an omission can be demonstrated through our children, our dearest treasures. Teenage suicides are mounting, and at an alarming rate. Whatever the specific reasons for this, the silent statement of those statistics makes it clear that for an increasing number of children, the living environment we adults are providing is sufficiently without hope to make suicide the only perceived alternative in too many cases. (One has to assume that drugs and premature sex—the more usual dead-end alternatives to dealing with hopelessness—have already been considered and, with some wisdom, been found wanting.)

There is a subliminal agreement between the adult world and its children that children must take what adults hand out. In its grisly aspects this agreement is evinced in corporeal abuse, sexual abuse, child pornography, etc. But on a more subtle level it occurs "naturally" in the world we provide for them. Children can only experience hope if this world is founded more on mutual respect and less on power games in which they, because of their youth, are ill-prepared to compete. Our unwillingness to provide a sanguine, nurturing environment is symptomatic of our indifference to their needs when they seemingly conflict with ours. As a small example, our wish to be entertained by sex and violence takes precedence over their needs to be reassured about the adult world they are being trained to enter. Children, like animals, are at our mercy—laboratory mice in a cage with a demented cat. We observe the phenomenon, take notes on stress, but have no wish to remove the cat. We continue to prefer self-aggrandizement to responsible concern.

*   *   *

We stand now at a crossroad. We will either annihilate this form of consciousness as we did at the times of Lemuria and Atlantis or we will survive because we change our truths to something which more nearly approaches fact. Those of us alive today have elected to be part of the big decision as to whether or not we will survive human willfulness. We are at this crossroad because, in good part, we have been kept in spiritual ignorance by intransigent religious institutions which were founded in the days of the spear and addressed to illiterate goat herders. Ignoring the growth of consciousness and the Intercontinental Ballistic Missile, religions have primarily concerned themselves, not with content, but with the preservation of their ancient forms. They continue, in spite of what we know about psychology, to manipulate the faithful through the destructive use of guilt and fear.

The message—have faith in God and he will reward you for that faith—was formulated long before Jesus. Its pragmatic benefit is that it keeps believers in attendance. Yet it should be clear by now that something is wrong either with the message or its delivery. Yet nothing changes. The faithful patiently wait upon a deliverance which will never come except as they do it for themselves. Thus week after week, as the millennia multiply, the faithful are asked to dwell upon their sins when they should be experiencing themselves as incarnate divinity.

We violate the law of the cosmos when we demean others or allow ourselves to be demeaned, *especially* in the name of piety. If we cannot validate/love ourselves, there is absolutely no chance that we can validate/love others. In spite of this, the common message coming from almost every pulpit is that humans are, in varying degrees, sinners. This is in direct contravention of cosmic law which demands a love of the One life in all of its manifestations. To declare that any part of that One should regard itself as unequal to any other (even to its God) is a terrible distortion. It is cosmically criminal. As the politician finds a national enemy to hate, so the minister finds, in the name of his God, a sinner to hate. Politically, we permit ourselves to rise up against our oppressors, but religions have rigged it so that no one dare rise up against the God of his oppressor priests. The devout then are left only with their guilt and no true understanding of where the path to salvation lies. (Salvation can only be achieved when there is a recognition of the *co*-divinity of All That Is. Co-divinity with God, the cosmos. Co-divinity with a molecule or cell, the cosmos.)

We have touched only briefly on matters of health. I urge you to read *As You Believe* if you have a greater curiosity in this subject or in solving other personal problems. For the purposes of this discussion, however, it can be said here that sickness is the result of in-turned punishment (guilt). Trained, in contradiction to cosmic law, to a sense of inadequacy, we become vulnerable in that persuasion. Vulnerable, we fall ill. Illness is but the symbolic outpicturing of personal disrespect. Yet again, it is the absence of an appreciation for the indivisible oneness of the cosmos which is at fault. Until we love *ourselves as we love our God,* we will manifest disease and illness along with other ills.

In summary, we are getting sicker, suffering more crime, and enduring a more toxic environment. The nuclear threat increases and so does the national debt—another symptom of our willingness to abuse children. This accelerating downward spiral can only end when there is a realignment of our direction to coincide with the laws of the cosmos as they are, not as we wish they were.

We must come to understand that we are Spirit incarnate if we are to find the high road. Further, we must tune the Spirit that is us to the laws of the cosmos. Getting ahead does not mean climbing to the top of the political, social, or economic heap. It means furthering the process of spiritual growth by assuming the responsibility for displaying the inner jewel that is us to its fullest glory and allowing others to do the same, for each jewel is different and the diamond cannot speak for the pearl.

I find great inspiration in the Creating Cosmos. It is the design of love—its dimensions, its possibilities, its demands, its expression, its meaning, its gifts. Love the Creating Cosmos for its awesome beauty and you will be loved in return for yours.

Index

About the Author

This is the place, should the length of the book permit, to credential the author with presumably illuminating remarks about his or her marital status, degrees, and public accomplishments. Nothing I have done, save a lifetime of thinking on the subject, can possibly credential me for this kind of book. It may be of interest, however, to pass along excerpts from a reading I had recently with Kevin Ryerson, a transmedium of some repute.

"Within the past lives context, your lives have been an exact chronological tracing of the flow of consciousness. First to Lemuria and then to Atlantis. You were at various times members of the priesthood in Egypt that dispensed those bodies of knowledge from the pyramid in the Gizeh complex to such areas as the Minoan Kingdom, the Ionian Kingdom upon Crete and throughout the various scattered Greek isles. Of course, some of your major successes were in the development of Zoroastrianism from Persia to Babylon when it was the center of influence. You were Melchizedek the Salem of Jerusalem some 5000 years past. The soul had incarnations in all historical regions which helped establish and distribute the teachings, particularly in such critical instances as the founding of the Alexandria Library. You as a soul travelled the regions of the Indus Valley as the bodies of information penetrated eastward. You had incarnations at the time of Buddha, in Nepal, amidst the Tibetans, the Taoists, the Confucists, and finally in the Zen orders of Japan.

"Note that many of these lifetimes were spent within monastic orders or mystery schools, as they were referred to. You were often the key to the verbal records—a verbal scribe. Bodies of knowledge would be imparted orally to thee, stored in the deep recesses of the right side of the brain and then brought up intuitively and recited to individuals as an appropriate revelation—this necessary to advance the more mundane or left brain sciences of the people."

Colophon

This book was designed by Michael Sykes and the author. It was typeset in Baskerville and produced by Michael Sykes, printed on acid-free paper and bound by Braun-Brumfield, Inc. in the Spring of 1985. The display type was composed by David Bunnett and the dust jacket was designed by the author. The illustrations are by the author.

The Theory of Laminated Spacetime   **Barbara Dewey**

The Theory of Laminated Spacetime (dis-continuous time) is surely the most revolutionary and far-reaching proposal ever suggested in the name of science. Not only does Dewey address the difficult subject of physics with a layman's good-humored clarity, but by extending the theory's physical implications she has discovered *the causes* of such phenomena as gravity, magnetism, polarity, elliptical orbits, electron shells, etc. Indeed, so many physical anomolies have been solved that the theory must inevitably be taken for fact. The book is therefore a must-read for every physicist and layman who has ever been curious about the laws under which the physical universe functions.

Hardcover, 116 pages, 31 diagrams   **$16.95 postpaid**

The Creating Cosmos   **Barbara Dewey**

Dewey's Theory of Laminated Spacetime is explored from the spiritual/humanistic point of view. Again, the theory may well be the single most important intellectual contribution ever offered in the name of understanding our universe, its purposes and origins, and its laws of cause and effect. With lively good humor the author examines contemporary truths concerning God and man and then goes on to offer a different truth. It is a truth which makes each of us "divine" creators in a cosmos where the act of creation itself is the sponsoring power. The book is for everyone who has ever wondered about the purpose of life, the God-concept, and the path to spiritual attunement.

Hardcover, 128 pages   **$16.95 postpaid**

As You Believe   **Barbara Dewey**

Because life itself is supported by the creating power of mind, each of us unavoidably produces the events of our lives. Knowing this we can now choose those events consciously. This is a practical how-to book for those who wish to stop pulling disasters into their lives. From cancer to money problems, from distressing job situations to unhappy love affairs, *As You Believe* presents a radically new and proven approach to personal problem solving. Based on the author's many years of experience as a psychic reader/healer, the book explains the inner processes which convert beliefs into events, details how to find the beliefs which are causing negative experiences, and then describes exactly how to effectively heal for positive "miracles."

Hardcover, 208 pages   **$18.95 postpaid**

Order Form

Please send me the following books postpaid:　　　Total

_______ Copies of *The Creating Cosmos* @ $16.95 each　　_______

_______ Copies of *The Theory of Laminated Spacetime*
　　　@ $16.95 each　　_______

_______ Copies of *As You Believe* @ $18.95 each　　_______

　　　Californians: Please add 6% sales tax　　_______

　　　　　Grand Total:　　_______

Name: _______________________________________

Address: _____________________________________

_______________________________ Zip: __________

Send to:

**Bartholomew Books
Box 634
Inverness, CA 94937
(415) 669-1664**